Correlation and Dependence

T0344594

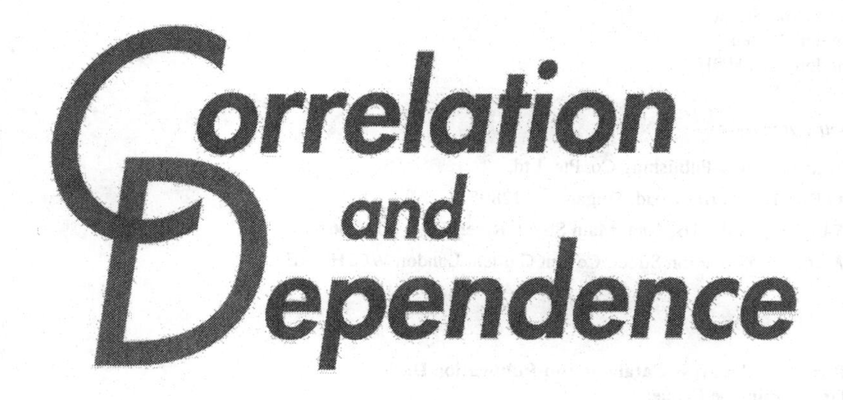

Correlation and Dependence

Dominique Drouet Mari
Université de Bretagne Sud, France

Samuel Kotz
George Washington University, USA

Imperial College Press

Published by

Imperial College Press
57 Shelton Street
Covent Garden
London WC2H 9HE

Distributed by

World Scientific Publishing Co. Pte. Ltd.
P O Box 128, Farrer Road, Singapore 912805
USA office: Suite 1B, 1060 Main Street, River Edge, NJ 07661
UK office: 57 Shelton Street, Covent Garden, London WC2H 9HE

Library of Congress Cataloging-in-Publication Data
Mari, Dominique Drouet.
Correlation and dependence / Dominique Drouet Mari, Samuel Kotz.
p. cm.
Includes bibliographical references and index.
ISBN 1-86094-264-4 (alk. paper)
1. Correlation (Statistics) 2. Dependece (Statistics) I. Kotz, Samuel. II. Title.
QA278.2 .M36 2001
519.5'37--dc21 2001016777

British Library Cataloguing-in-Publication Data
A catalogue record for this book is available from the British Library.

Copyright © 2001 by Imperial College Press
All rights reserved. This book, or parts thereof, may not be reproduced in any form or by any means, electronic or mechanical, including photocopying, recording or any information storage and retrieval system now known or to be invented, without written permission from the Publisher.

For photocopying of material in this volume, please pay a copying fee through the Copyright Clearance Center, Inc., 222 Rosewood Drive, Danvers, MA 01923, USA. In this case permission to photocopy is not required from the publisher.

Printed in Singapore.

to Rosalie,
and to Marguerite.

Preface

The concept of dependence permeates our Earth and its inhabitants in a most profound manner. Examples of interdependent meteorological phenomena in nature, interdependence in medical, social, and political aspects of our existence, not to mention economic structures, are too numerous to be cited individually. Moreover, the dependence is obviously not deterministic but of a stochastic nature.

It is therefore somewhat surprising that the concepts and measures of dependence did not receive sufficient attention in the statistical literature, at least until as late as 1966 when the pioneering paper by E.L. Lehmann has appeared. The concept of *correlation* (and its modifications) introduced by F. Galton in 1885 dominated statistics during some 70 years of the 20-th century, practically serving as the *only* measure of dependence, often resulting in somewhat misleading conclusions. The last thirty years of the 20-th century have witnessed a rapid resurgence in investigations of dependence properties from statistical and probabilistic points of view but the first -and to the best of our knowledge- the only text (of some 400 pages) devoted to dependence concepts (by Harry Joe) appeared as late as 1997. Moreover it seems to us that no Department of Statistics (or/and Mathematics) in the U.S.A and Europe offer courses dealing specifically with *dependence concepts and measures.*

Our monograph should thus be viewed as an initial attempt to help to remedy the situation, and we have written it for a graduate course or a seminar covering correlation and dependence concepts and measures. An initial background in mathematical statistics and probability theory and integral calculus is required. A few notions related to stochastic processes are used in Chapter 6.

This monograph is not a full-scale expedition up another statistical Alp. Rather it is a tour over a somewhat neglected but important terrian. Chapter 1 introduces the notation and basic definitions. Chapter 2 deals with the concept of correlation. Historical background is traced and differences between correlation and other types of probabilistic dependence are examined. This chapter can be read and understood by anyone with most limited preparations. Chapters 3 and 6 deal with concepts and measures of dependence, respectively. We have examined in these chapters numerous *measures* of dependence proposed in the literature, especially in the field of survival analysis, and their connection to various *concepts* of dependence. We have been able to collect materials from a large number of sources and have provided a comprehensive and unified discussion which is not available in monographic literature. These two chapters constitute the core of the monograph. Chapter 4 deals with a popular concept of copula : a focused expression of dependence between two (or several) random variables, totally stripped of any other characteristics. This concept is useful as a tool for modelling of bi- or multivariate distributions. A monograph by R. Nelsen on this subject has appeared in 1998. However the overlap between our chapter and the monograph is not substantial, since we have emphasized post-1997 results as well as topics that are not discussed in Nelsen's contribution.

The fifth chapter deals with a family of distributions which -in our opinion- represents a natural and general method of generating dependence between random variables applicable to various physical models. Much of the material in this chapter is based on the research of Samuel Kotz.

It is our pleasure to acknowledge the support and encouragement from the Chairman of the Imperial College Press, Dr. K.K. Phua, and the Editor, Ms. E.H. Chionh. Our special thanks are extended to Dr. A. Müller (University of Karlsruhe, Germany) who provided us with most valuable comments and suggestions, to Professors N.L. Johnson (Chapel Hill, U.S.A.) and I.G. Bairamov (Ankara, Turkey) for their assistance related to the text of Chapter 2 and 5, respectively, and to Mrs. Françoise Nadot (Université de Bretagne Sud) for her help in locating and collecting materials.

Contents

Correlation and Dependence

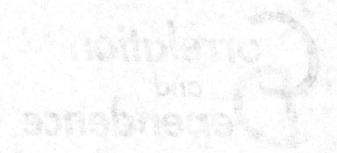

Chapter 1

Notations and Definitions

1.1 Notations

(1) $C^m(E)$ is the space of the real valued functions, defined on E with a continuous m-th derivative.
(2) $f \circ g(x)$ denotes the composition of the functions f and g : $f \circ g(x) = f[g(x)]$,
(3) in R^2, D_1 and D_2 denote $\partial/\partial x$ and $\partial/\partial y$ respectively, in R^n, D_j denotes $\partial/\partial x_j$.
(4) Bold lower-case letters are used for vectors, for example $\mathbf{x} = (x_1, ..., x_n)$.
(5) Random variables are denoted by upper-case letters X, Y.
(6) Random vectors are denoted with bold letters $\mathbf{X}$, $\mathbf{Y}$.
(7) For two vectors $\mathbf{x}$ and $\mathbf{y}$, the operators $\vee$ and $\wedge$ denote,respectively, componentwise maximum and componentwise minimum, i.e. $\mathbf{x} \vee \mathbf{y} = (max(x_1, y_1), ..., max(x_n, y_n))$ and $\mathbf{x} \wedge \mathbf{y} = (min(x_1, y_1), ..., min(x_n, y_n))$.
(8) The transpose of a vector $\mathbf{X}$ is denoted by $\mathbf{X}'$.
(9) $\mathbf{e} = (1, 1, ..., 1)'$.
(10) I is the identity matrix.
(11) The Kronecker product $\otimes$ is used for product of matrices.
(12) $\delta(u)$ is the Dirac measure at point u.
(13) $\sharp$ is used for the cardinality of a set.
(14) $\emptyset$ denotes the empty set.
(15) Greek lower-case letters are used for parameters of families of distribution such as α, θ.
τ designates the coefficient of Kendall, ρ the linear correlation coefficient, ρ_S the Spearman coefficient.

(16) $F(x,y)$ or $F_{XY}(x,y)$ denotes a bivariate cumulative distribution function (c.d.f.); its marginals are denoted by $F_1(x)$ and $F_2(y)$. $F(\mathbf{x})$ or $F_{\mathbf{X}}(\mathbf{x})$ corresponds to a multivariate cumulative distribution.
(17) $C(u,v)$ denotes a copula, that is the c.d.f. on the unit square with uniform marginals.
(18) $S(x,y)$ or $\overline{F}(x,y)$ represents a bivariate survival function.
(19) Its marginals are denoted by $S_1(x)$ and $S_2(y)$.
(20) We suppose, in general, that the bivariate probability density function (p.d.f) $f(x,y)$ ($c(u,v)$ for the copula) exists.
(21) Marginal densities are denoted by $f_1(x)$ and $f_2(y)$.
(22) Bivariate hazard function is denoted by $h(x,y) = \frac{f(x,y)}{S(x,y)}$, and can be viewed as the conditional density of (X,Y) at the point (x,y) given that $(X > x, Y > y)$.
(23) Bivariate cumulative hazard is denoted by $H(x,y) = \int h(u,v)dudv = -Log(S(x,y))$.
(24) $\mathcal{F}(F_1,F_2)$ denotes the family of the bivariate distributions with marginals F_1 and F_2.
(25) In this family, the upper and lower Fréchet bounds are denoted respectively by F^+ and F^-, F^0 denotes the independent distribution. On the unit square, the corresponding notations are C^+, C^- and C^0.
(26) $N(\mu,\sigma^2)$ corresponds to the univariate normal distribution, with mean μ and variance σ^2. When $\mu = 0$ and $\sigma^2 = 1$ we have the standard normal distribution. $N(\mu,\Sigma)$ corresponds to the multivariate normal distribution with mean vector μ and covariance matrix Σ.
(27) $\Gamma(\nu,\lambda)$ is used for the gamma distribution with shape parameter ν, and scale parameter λ.
(28) $Beta(a,b)$ is used for the Beta distribution.
(29) $U(a,b)$ is used for the uniform distribution on the segment $[a,b]$.
(30) The cumulative distribution function of the normal standard distribution is denoted Φ , Φ^{-1} is its inverse function or quantile.
(31) $\sim$ means *distributed as*
(32) $\stackrel{L}{=}$ means *equal in distribution.*
(33) $\nearrow$ means *non decreasing,* $\searrow$ means *non-increasing.*
(34) $\ll$ is used for partial ordering between two distributions. The subscript stipulates the type of ordering.

1.2 Definitions

(1) Conditional hazards
In the case of survival variables, the connection between (X, Y) can be studied using the conditional hazard $h(y/X = x) = \frac{f(y/X=x)}{S(y/X=x)}$, that is the hazard of Y given X failed at x, also denoted by $h_{2/1}(x, y)$,

(2) or using $h(y/X > x)$, the hazard of Y given X surviving beyond x (or is censoring at x), denoted by $h_2(x, y)$. Reversing the roles of x and y, one also defines $h_{1/2}$ and h_1. We have :

$$h(x,y) = \frac{D_1 D_2 S}{S},$$

$$h_{2/1} = \frac{-D_1 D_2 S}{D_1 S},$$

$$h_2 = \frac{-D_2 S}{S}. \tag{1.1}$$

If $L = LogS(x, y)$ and $l = D_1 D_2 L$ then :

$$\begin{aligned} l &= \frac{D_1 D_2 S}{S} - \frac{D_1 S}{S} \cdot \frac{D_2 S}{S} \\ &= h - h_1 h_2. \end{aligned} \tag{1.2}$$

Also :

$$l = -D_1 h_2 = -D_2 h_1 \tag{1.3}$$

and

$$h_{2/1} - h_2 = -\frac{D_2 h_1}{h_1}. \tag{1.4}$$

(3) The expectation of a random variable X with the cumulative distribution function F is $E(X) = \int x f(x) dx$ if F is absolutely continuous with density f and is denoted by $\int x dF = \Sigma x p(x)$ if the distribution is concentrated over points x with masses $p(x)$. If g is a real-valued function, then we define the expectation of the random variable $g(X)$ by $E(g(X)) = \int g(x) dF$.

(4) Expected residual life is defined by $m(x) = E(X - x/X > x)$.

(5) Laplace transform for a positive random variable X with cumulative distribution function F is denoted by $\varphi^{-1}(t) = E(e^{-tX})$.

(6) Convex set:
A set A in R^n is convex if for all $\mathbf{x} \in A$ and for all $\mathbf{y} \in A$, $\lambda\mathbf{x} + (1-\lambda)\mathbf{y} \in A$, for all $0 < \lambda < 1$.

(7) Convex function
A real valued function g defined on A, an open convex set of R^n is called convex if $g\,(\lambda\mathbf{x} + (1-\lambda)\mathbf{y}) \leq \lambda g(\mathbf{x}) + (1-\lambda)g(\mathbf{y})$ for all $x,\ y \in A,\ 0 < \lambda < 1$. A function is componentwise convex if it is convex in each component, the other components being held fixed.

(8) Supermodular function
A function $\Phi\ : R^m \mapsto R$ is supermodular if

$$\Phi(x \vee y) + \Phi(x \wedge y) \geq \Phi(x) + \Phi(y).$$

A necessary and sufficient condition for Φ to be superadditive is that Φ is superadditive in (x_i, x_j), the other variables being held fixed. It is easy to prove that, if Φ possesses second partial derivatives, then the superadditivity is equivalent to the positivity of $\frac{\partial^2\Phi}{\partial x_i \partial x_j}$ for all i and $j,\ j \neq i$. See Marshall and Olkin [152] for more details and examples. A supermodularfunction is sometimes referred to as a superadditive, L-superadditive or quasi-monotone functions.

(9) Total positivity (TP) :
This concept has been developed by Karlin[123]. A real-valued function $K(x, y)$ of two variables ranging over linearly ordered sets X and Y is said totally positive of order r (TP_r) iff for all

$$x_1 < x_2 < \ldots < x_m\ ,\ y_1 < y_2 < \ldots < y_m\ ; x_i \in X,\ y_i \in Y\ ;\ 1 \leq m \leq r$$

we have

$$K\begin{pmatrix} x_1, x_2, \ldots x_m \\ y_1, y_2, \ldots, y_m \end{pmatrix} = \begin{vmatrix} K(x_1,y_1) & K(x_1,y_2) & \ldots & K(x_1,y_m) \\ K(x_2,y_1) & K(x_1,y_2) & \ldots & K(x_2,y_m) \\ . & . & \ldots & . \\ . & . & \ldots & . \\ . & . & \ldots & . \\ K(x_m,y_1) & K(x_m,y_2) & \ldots & K(x_m,y_m) \end{vmatrix}$$

If this result holds for all $r \geq 1$, the function is said to be totally positive of order infinity (TP_∞). We shall use mainly the concept of TP_2, or the reverse concept of RR_2 (when the sign of the determinant is reversed).

(10) Pólya function of order two
If a real valued function f satisfies $f(x-y) = K(x,y)$, where K is TP_2, the function is said to be Pólya function of order two. In that case $-log(f)$ is convex. This property is used in case of product convolution.

(11) Random exchangeable vector
A random vector $\mathbf{X} = (X_1, X_2, ..., X_n)$ with distribution function F is exchangeable if the distribution is kept by permutation : $(X_1, X_2, ..., X_n) \stackrel{L}{=} (X_{\pi_1}, X_{\pi_2}, ..., X_{\pi_n})$, where π is any permutation of $\{1, 2, ..., n\}$, and all the marginals are the same.

Chapter 2

Correlation and Dependence : An Introspection

2.1 Independence

Verbal definitions of independence appeared occasionally in writing of various authors in the eighteenth and nineteenth centuries (over 100 years before F. Galton's preoccupation with correlation). T. Bayes in his 1763 paper is concerned with the notion of independence and defines it as follows:

> "Events are independent when the happening of any one of them does neither increase nor abate the probability of the rest [p. 376]."

Apparently Bayes does not distinguish between independence and pairwise independence.

A similar definition is provided in 1767 by R. Price, a close friend and contemporary of T. Bayes, in his volume *Four Dissertations* (republished in the second edition with additions by A. Miller and T. Casell in London, 1768). He states in a footnote [p. 440] :

> "Two events are independent when the happening of one of them has no influence on the other."

Next [p. 452] he adds the following rather curious proposition :

> Proposition 2nd : The *improbability* that two independent events, each of them not improbable, should both happen, cannot be greater than the odds of *three* to *one* ; this

> being the odds that two equal chances shall not both happen; and an equal chance being the lowest event of which it can be said that it is not improbable.

A British mathematician W. Emerson (see I. Hacking (1975) for details about Emerson), in his 1776 treatise entitled *Miscellaneous Treatise containing several Mathematical Subjects*, in the Article 1, The Laws of Chance [pp. 1-46], provides the following two definitions :

> Def. V. Events are *independent* when they have no manner of connection with one another; or when the happening of one neither forwards nor obstructs the happening of any other of them.
>
> Def. VI. An event is *dependent* when the probability of its happening is altered by the happening of some other.

P.S. Laplace in the second book of his classical *Théorie Analytiques des Probabilités* (1812) presents perhaps the earliest mathematical description of independence pointing out that if $\{E_i\}$ is a sequence of independent simple events with $p_i = P(E_i)$, then $P(E_1, ..E_n) = \Pi_{i=1}^n p_i$ and then notes that in the case of two simple events where the supposition of the occurrence of the first E_i affects the probability of the occurence of the second E_2 we have $P(E_1E_2) = P(E_2/E_1)P(E_1)$.

G. Boole's attempts in his famous and influential *Laws of Thought* (1854) to define independence in a fascinating story. In the words of J. M. Keynes (1921), "He first wins the reader's acquiescence by giving a perfectly correct definition in his *Laws of Thought* (p. 255)" namely :

> "Two events are said independent when the probability of the happening of either of them is unaffected by our expectation of the occurrence of failure of the other."

His second definition, however, claims a quite different assertion :

> "We must regard the events as independent unless we are told either that they must occur or that they cannot occur."

In other words the events are independent unless we know for certain there is in fact an invariable connection between them. Or using Boole's terminology : "Given simple events x and z, as long as xz is possible, x and z are independent." Keynes (1921) discusses in some detail paradox arising due to employing the term of independence in a double sense. This mistake was pointed out almost immediately by H. Wilbraham in *Phil. Mag*, 4th series , Vol VII (1854) while reviewing Boole's masterpiece which, however, resulted in a hot rebuttal from Boole. There were other rather feeble attempts in the literature (e.g. by H. McColl in his sixth paper on the "Calculus of Equivalent Statements", 1897, *Proc Lond. Math. Soc.*, Vol **28**, p. 556) to pinpoint the discrepancy of Boole's definitions until Keynes (1924) clarified the situation in his classical *Treatise on Probability Theory.* Moreover the second definition of independence given by Boole was also adapted in A. Macfarlane's "Principles of the Algebra of Logic " (paper in Mathematical Questions from the *Journal of Education*, Vols **32** and **36**).

A. De Morgan's (1806-1871) definition of independence in his *Essay on Probabilities* (1838) is almost identical with Boole's second definition : two events are independent if the latter might have existed without the former or the former without the latter, for anything that we know for the contrary.

Keynes's lucid example of the danger of wrongful assumption of independence is worth quoting almost verbatim.

Consider the problem of determining the probability of throwing two heads twice in two consecutive tosses of a coin.

> "The plain man generally assumes without hesitation that the chance is $(\frac{1}{2})^2$. For the *a priori* chance of heads at the first toss is $\frac{1}{2}$, and we might naturally suppose that the two events are independent - since the mere fact of heads having appeared once can have no influence on the next toss. But this is not the case unless we know for certain that the coin is free from bias. If we do not know whether there is bias, or which way the bias lies, then it is reasonable to put the probability somewhat higher than $(\frac{1}{2})^2$. The *fact* of heads having appeared at the first toss is not the cause of head appearing at the second also, but the *knowledge*, that the coin has fallen heads already, affects our forecast of its falling thus in the future, since heads in the past may have been due to a cause which will favor heads in the future. The possibility of bias in a coin, it may be noticed, always favors 'runs' ; this possibility increases the probability both

of 'runs' of heads and of runs of 'tails'."

Keynes's definition of independence is worth noting. The definition is based on the idea of representing *probability* by the symbol a/h where h is the premise(s) of an argument and a is its conclusion; and a/h designates the argument from h to a as well as -dropping the prefix P- the probability (the degree of rational belief about a to which the argument authorizes). In Keynes's opinion the introduction of the symbol a/h is an essential step. His definition of independence is "the familiar statement" : "If $a_1/a_2 = a_1/h$ and $a_2/a_1h = a_2h$, then a_1/h and a_2/h are independent."

Mosteller and Tukey (1977) emphasize :

> "We must be clearer about the abused word "dependence" and its relatives. When we say "y depends on x", sometimes we intend *exclusive dependence* , meaning that, if x is given, then the value of y follows, usually because the existence of a law is implied. In mathematics if y is the area of a circle and r the radius, then $y = \pi r^2$ illustrates this reguar type of *exclusive dependence.*
>
> At other times "y depends on x" means *failure of independence*, usually in the sense of "other things being equal" as in "the temperature of the hot water depends on how far the faucet is from the heater." Clearly, it may depend of other things, such as the setting of the heater and the building's temperature, to say nothing of whether the long pipe between the faucet and the heater is on a cold outside wall or a warm inside one. These are two quite different ideas of dependence, and the use of one word for both has often led to troublesome, if not dangerous, confusion.
>
> Then there are the mathematical usages "dependent variable" and "independent variable". These have been extremely effective in producing confusion when dealing with data."

2.2 Zero Correlation Versus Dependence

Correlation is among the basis concepts in statistics, familiar to researchers in natural, social and medical sciences, not to mention economics. Section

2.4 contains a brief summary of the historical development of the concept of correlation and its uses. For the present we are concerned mainly with its status, usage and interpretation, merely noting that the concepts developed in the nineteenth century, steadily widened its appeal to research workers in the twentieth century and now, in the twenty-first century, most laymen have at least a rudimentary idea of the meaning of "correlation". Among the latter, however, few know the names of its foremost creators.

In mathematics, especially in advanced calculus and measure theory, the notion of "independence"-introduced, presumably by Henri Lebesgue and/or his contemporaries- is, of course, fundamental. Probability theory, which developed as a branch of measure theory with an eye to applications in the real world, transmitted this concept to statistical methodology. We thus have two concepts –*independence* (and its negation *dependence*) and *correlation* (and its counterpart, *lack of correlation*) prominently utilized in statistical sciences. In their attempts to " simplify" statistical methodology (catering for a supposed common lower ground) many authors have not properly distinguished between independence and lack of correlation (often referred to "zero correlation" or "uncorrelatedness"). Consequently casual readers of these writings are often under the impression that to establish practical independence (or, more specifically, absence of meaningful relationships among the variables) it suffices to verify that the correlation coefficients are, effectively zero. There has been considerable degree of harm caused by this attitude in various contexts - most prominently in medical and social sciences applications- by confusing these concepts and deducing incorrect and sometimes damaging conclusions. A typical quote from a recent newspaper article reads : "Correlations do arrest attention to a strong relationship".

2.2.1 *Linear relationship*

There are now numerous books and journal articles- notably the book by Darral Huff "How to Lie with Statistics" –which draw attention to these problems. It is, by now, well-established that correlation coefficients measure the degree of *straight line* or *linear* relationship only, and that there are situations in which correlations are zero but nevertheless, strong nonlinear relationships exist among variables, which are indeed highly *dependent* in a probabilistic, as well as intuitive senses.

Some initial verbal descriptions of correlation in the early twentieth century tended to obscure this important limitation of "correlation" as a measure of dependence. Witness, for example, the explanation in a well-known and authoritative textbook by Brown [33] :

> " Correlation may briefly defined as a tendency towards concomitant variation and the so-called correlation coefficient is simply a measure of such tendency, *more or less adequate according to the circumstances of the case.* (Italics are from the present authors !)"

Indeed, numerous attempts were made by earlier (pre–World War I) writers to apply correlation coefficients indiscriminately to miscellaneous material. This was, of course, an abuse of this method of measuring relationship.

This activity was criticized by G. Yule as early as 1912 in his classical *An Introduction to the Theory of Statistics* (a direct predecessor of the present-day authoritative and popular *Kendall's Library of Statistics*) wherein the author emphatically points out that correlation may be used strictly as a *measure* of relationship *only* when such relationship has been determined by other investigation(s), to follow a straight–line relationship. Yule emphasized that the correlation coefficient is, even then, "merely a statement" of the position of the straight–line of closest fit on a chart where the units are the standard deviations of the variables as this position is determined by the least square adjustment". Consequently this coefficient should never be used without first investigating the relationship thoroughly enough to see if it follows a straight line. This restriction was often ignored by some prestigious statisticians at least in the early years of the twentieth century.

We note that even A. Bowley- a prominent British statistician in the first decade of the twentieth century- whose textbook, *Elements of Statistics* [30], went through fourteen editions (up to 1936), asserted that "correlation coefficient tends to be the ratio of the number of causes common to the genesis of two variables to the whole number of individual causes on which each depends."

One of the earliest examples of a situation wherein the correlation coefficient is near zero, although there is a strong relationship between the two variables, was provided by W.G. Reed (1918). He took records of the height of the higher high water at Old Point Comfort, Virginia, U.S.A. and

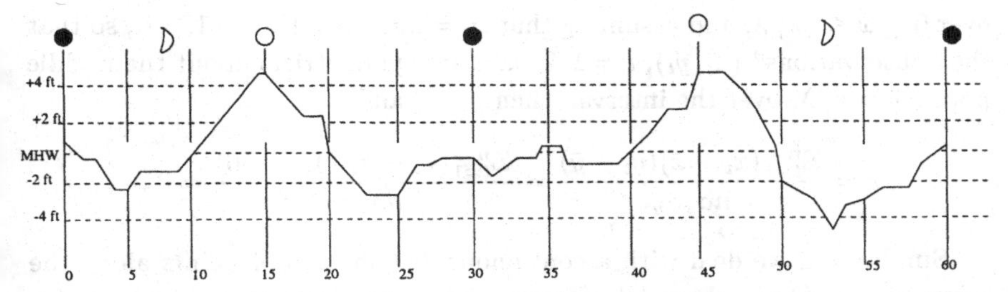

Fig. 2.1 DAYS AFTER NEW MOON - JULY 29, 1916. *Predicted height of the higher high water for each day after new moon : with reference to mean high water (MHW) at Old Point Comfort VA (U.S. Coast and Geodesic Survey General Tide Tables for the year 1916, p.103).*

the time after new moon for the period of 61 days originated at new moon on July 29, 1916 (presented in Figure 2.1). In an obvious notation, the correlation coefficient was calculated to be :

$$r = -0.106\,(\pm 0.088)\,,$$

where ± 0.088 is the estimated probable error :

$$E_r = \pm 0.674(1 - r^2)/\sqrt{n} \quad (n = 61)\,.$$

From Figure 2.1, it is clear that there is a harmonic relation between the variables , while the correlation coefficient is nearly zero, because the straight line of closest fit is nearly parallel to the horizontal axis (yielding the angle between the line and the axis to be nearly zero).

2.2.2 *Non-linear relationship*

Prompted by Reed's remarks, a well known American mathematical statistician, H.L. Rietz (1918) was the first, to the best of our knowledge, to study the possible relations between two variables when their correlation coefficient is zero. He emphasized limitations on the generality of the correlation coefficient as a measure of "correlation". Perhaps having in mind Reed's example, he noted that if we observe as corresponding variates the coordinates in a simple harmonic motion:

$$y = cos(\lambda x)$$

over $0 \leq x \leq 2\pi/\lambda$, and assuming that $x_t = 2\pi t/(n+1)$, $t = 1, .., n$, so that the "observations" (x_t, y_t), $t = 1, 2, ..., n$ are symmetrical about the middle point, $\bar{x} = \pi/\lambda$, over the interval, then $\bar{y} = 0$ and

$$r = \frac{\Sigma_{t=1}^{n}(x_t - \bar{x})(y_t - \bar{y})}{n\sigma_x\sigma_y} = \frac{\Sigma_{t=1}^{n}(x_t - \pi/\lambda)(y_t - 0)}{n\sigma_x\sigma_y} = 0 \, .$$

Similarly, if we deal with a continuous distribution of points along the curve $y = cos(\lambda x)$, where kdx (k constant) is the frequency of values of x in the interval dx, we have, as simple calculations show :

$$r = \frac{k}{s^2} \int_0^{2\pi/\lambda} y(x - \frac{\pi}{\lambda})dx \, ,$$

where s^2 is the geometric mean of the second moments of the distribution about the two lines $x = \pi/\lambda$ and $y = 0$. In particular, we obtain $r = 0$ when there is the relationship :

$$y = cos(\lambda x) \, .$$

If the points were distributed uniformly over intervals of width Δx in a thin band along the curve $y = cos(\lambda x)$, instead of being exactly on the curve, r would be *nearly* zero.

Indeed, about 100 years ago, early writers sensed that "if the regression curve is of a certain shape, the value of r will be very small, even though practically perfect correlation exists", (Davenport 1918 [54]). In Yule's words : Here a horizontal regression line indicates absence of correlation, while the decreasing scedastic curve indicates dependence.

H.L. Rietz (1918) strengthened this statement by showing that "if the regression curve is of a certain shape, the value of r would be zero in certain cases, even when the one variable is a certain trigonometric function of the other". He notes that instead of $y = cos(\lambda x)$ we may take a more general relationship:

$$y = f(x)$$

over the interval $(-a, a)$, where $f(x)$ is *any* single-valued function, symmetrical about $x = 0$. If the distribution of the points among the intervals of equal length Δx is "uniform", in the sense just described above for the case $y = cos(\lambda x)$, we have more generally that $r = 0$ if $y = f(x)$. The condition

Table 2.1 *Number of deaths at birth in England and Wales, years : 1881-1890*

umber deaths birth er 1000 orn	Total number of births during the decade 1500-2500	3500-4000	4500-5000	10000-15000	15000-20000	30000-50000	50000-90000
5-2.0	-	-	2	-	-	-	-
0-2.5	1	-	1	1	-	-	-
5-3.0	1	3	1	-	-	-	-
0-3.5	1	5	2	4	-	1	2
5-4.0	5	6	5	8	5	5	9
0-4.5	6	5	8	23	4	9	6
5-5.0	2	5	9	14	11	7	5
0-5.5	7	3	6	14	6	8	7
5-6.0	5	3	4	5	2	5	4
0-6.5	1	5	1	-	4	1	1
5-7.0	3	1	1	3	-	2	1
0-7.5	1	1	-	-	-	4	-
0-8.5	-	-	-	-	-	-	-
5-8.0	-	-	-	-	-	1	-
5-9.0	1	1	-	-	1	-	-
0-9.5	-	-	-	-	-	-	-
5-10.0	1	-	-	1	-	-	-
).0-10.5	-	-	-	-	-	-	-
).5-11.0	1	-	-	-	-	-	-
tal	36	38	40	73	33	43	35
verages	5.29	4.71	4.45	4.68	4.99	5.13	4.64
andard evia-on	1.77	1.37	1.09	1.01	0.99	1.12	0.87

of symmetry on $f(x)$ over $(-a, a)$ implies that :

$$\int_{-a}^{a} xy dx = \int_{-a}^{a} x f(x) dx = 0 .$$

A great variety of simple functions satisfy this condition.

It is important to remember that the use of r *cannot* lead to indication of a greater degree of dependence than it actually exists, but in cases of nonlinear regression it may lead to inferring a *smaller* degree of dependence that actually exists.

(However, in the special case when random variables X and Y have a joint bivariate normal distribution (not just normal marginals) the zero correlation *does* imply that X and Y are independent.)

In the second edition of his *An Introduction to Mathematical Statistics* (1912) (p. 163) G.U. Yule provided the Table of the numbers of births and the death rate per 1000 of deaths at birth (Table 2.1).

Although the average death rate for the seven groupings of numbers of births does not appear to vary in a systematic way and the correlation coefficient is near zero, there does appear to be an increase in standard deviation of death rates as the total number of births decreases, indicating dependence between the two variables.

2.2.3 *A technical discussion*

We now proceed to a slightly more technical discussion of relationships between zero correlation ("uncorrelatedness") and dependence (or equivalently, independence).

If the covariance of two non-degenerate random variables X and Y :

$$cov(X, Y) = E(XY) - E(X)E(Y)$$

is zero, i.e.;

$$E(X)E(Y) = E(XY), \tag{2.1}$$

the variables are said *uncorrelated.*

If the two variables are *independent*, i.e.

$$F_X(x)F_Y(y) = F_{X,Y}(x, y) \qquad \forall\, x,\ \forall\, y \tag{2.2}$$

or, if derivatives exist, equivalently,

$$f_X(x)f_Y(y) = f_{X,Y}(x, y) \qquad \forall\, x,\ \forall\, y \tag{2.3}$$

then they are also uncorrelated. For discrete random variables, independence implies that :

$$P(X = a/Y = b) = P(X = a)$$

for all a and b in the support of the distributions.

However, variables which are not independent–that is they are *dependent*– can also be uncorrelated. This will be so, for example, if

$$E(X/Y = y) = E(X),\ \forall y \tag{2.4}$$

or

$$E(Y/X = x) = E(Y),\ \forall x \tag{2.5}$$

(or, of course, both).

Clearly, independence is a much more stringent requirement than uncorrelatedness. Conditions (2.2) and (2.3) require equality of function *for all x and y*, while condition (2.1) requires equality of a function of expected values and simple product-moment of X and Y. A *rough* analogy to this comparison is to note that the arithmetic means of two sets of numbers can be equal without necessarily implying that each number in one set is equal to the corresponding number in the second set.

A very simple example of two dependent but uncorrelated variables is given by the joint discrete distribution (taking on three values only) :

$$P(X = Y = 0) = P(X = Y = 1) = P(X = -1, Y = 1) = \frac{1}{3}.$$

Here the marginal distributions are

$$P(X = -1) = P(X = 0) = P(X = 1) = \frac{1}{3}\ ;$$

$$P(Y = 0) = \frac{1}{3};\ P(Y = 1) = \frac{2}{3}\ .$$

Hence :

$$E(X) = \frac{1}{3}(-1 + 0 + 1)\ ;\ E(Y) = \frac{1}{3} \times 0 + \frac{2}{3} \times 1 = \frac{2}{3}\ ,$$

and

$$E(XY) = \frac{1}{3}(0 \times 0 + 1 \times 1 + (-1 \times 1)) = 0.$$

So $cov(X, Y) = 0.\frac{2}{3} - 0 = 0$, i.e. X and Y are uncorrelated, but they are clearly dependent since $P(X = 1/Y = 0) = 0$ but $P(X = 1/Y = 1) = \frac{1}{2}$.

In this case, we note that $E(X/Y = y)$ does not depend on y :

$$E(X/Y = 0) = E(X/Y = 1) = 0.$$

More generally, the same result will hold if

$$P(X = Y = 0) = P(X = Y = \theta) = P(X = -\theta, y = \theta) = \frac{1}{3}$$

for any real θ.

2.3 Some Geometrical Examples

A number of interesting examples can be constructed by considering simple continuous distributions on the unit square $[0, 1] \times [0, 1]$. The uniform distribution over this square has joint density function

$$f_{X,Y}(x, y) = 1, \ \ 0 \le x \le 1, \ 0 \le y \le 1. \tag{2.6}$$

Clearly, X and Y are uncorrelated *and* independent.

We now describe several modifications of this density producing distributions which retain the property of uncorrelatedness but lose the property of independence. All of these cases are constructed by changing the density (Eq. (2.6)) to 2 or zero over various portions of the unit square. By ensuring that the total area with density 2 is equal to the total area with zero density, we still retain a proper distribution with density integrating to 1 over the unit square.

Two such modified densities are represented diagrammatically in Fig. 2a-b. In the cross-hatched areas the density is still 1; in the black areas it is 2; in the white area it is zero. These figures are obtained by inserting one of two designs -*nicks* or *notches.*

A *nick* is composed of a white rectangle with two symmetrical black borders. Figure 2.2-a shows a nick inserted in a unit square, with its base

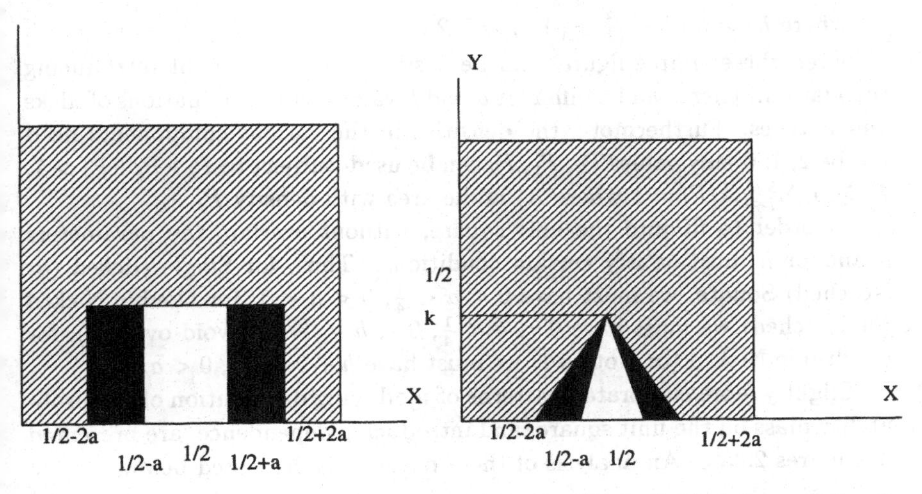

Fig. 2.2 *2a : Single Nicked Square, $0 \leq a \leq 1/4$, 2b : Single Notched Square $0 \leq a \leq 1/4$.*

on the X-axis. Since the center of the base is at $X = \frac{1}{2}$, it is easy to see that

$$E(X/Y = y) = \frac{1}{2}, \ \forall y, \ 0 \leq y \leq 1,$$

so that uncorrelatedness is preserved. But independence is clearly *not* preserved. For example,

$$E(Y/x) = \begin{cases} \frac{1}{2} & \text{for } |x - \frac{1}{2}| > 2a \\ \frac{1}{2}(1 + h^2)(1 + h)^{-1} & \text{for } a < |x - \frac{1}{2}| < 2a \\ \frac{1}{2}(1 + h) & \text{for } |x - \frac{1}{2}| < h. \end{cases}$$

Hence the distribution of Y *does* depend on the value of x. A single nicked square distribution was introduced by Borkowf *et al.* [27].

A *notch* is composed of two isosceles triangles, one superposed on the other, as in Figure 2-b - a single notched square. Here, again it is easy to see that Eqs.(2.4) and (2.5) hold, so that X and Y are uncorrelated, but dependent.

$$E(Y/x) = \begin{cases} \frac{1}{2} & \text{for } |x - \frac{1}{2}| > 2a \\ \frac{1}{2}(1 + h_1^2)(1 + h_1)^{-1} & \text{for } a < |x - \frac{1}{2}| < 2a \\ \frac{1}{2}(1 + h_1^2 - 2h_2^2)(1 + h_1 - 2h_2)^{-1} & \text{for } 0 < |x - \frac{1}{2}| < a, \end{cases}$$

where $h_j = h\left(1 - \frac{|x-\frac{1}{2}|}{(3-ja)}\right) \quad j = 1, 2$.

Even these simple figures can be easily extended without introducing correlation. There can be different a and h values and combinations of nicks and notches. Furthermore, the densities in the various components need not be 2, 1, 0 -any values P_2, P_1, P_0 can be used- subject to $P_2+P_1+P_0 > 1$, $P_j \geq 0$, $\Sigma_{j=1}^{3} A_j P_j = 1$ where A_j is the area with density P_j.

In order to fit into the unit square, without overlap, the parameters a and/or h must satisfy certain conditions. Thus, for Single Nicked (or Notched) Square, we must have $0 < a < \frac{1}{4}$, $0 < h < 1$; for Double Nicked (or Notched) we must have $0 < a < \frac{1}{4}$, $0 < h \leq \frac{1}{2}$ (to avoid overlap); for quadruple Nicked (or Notched) we must have $h + 2a < \frac{1}{2}$, $0 < a$.

Slightly more elaborated patterns of modifying distribution of the probability mass on the unit square, and introducing dependence, are presented in Figures 2.3a-c. An analysis of these patterns is presented below :

(1) Original pattern (Figure 2.3a) :

$$E(X/Y = y) = \begin{cases} \frac{1}{2} & \text{for } c < y < 1 - c \\ \frac{1}{2} + c - c^2 & \text{for } 0 < y < c \\ \frac{1}{2} - c + c^2 & \text{for } 1 - c < y < 1 \,. \end{cases}$$

Indeed, for $0 < y < c$, we have $(0 \times c \times \frac{1}{2}) + (1 \times (1 - 2c)\frac{1}{2}) + (2 \times c \times (1 - \frac{1}{2}c)) = \frac{1}{2} + c - c^2$.

Both X and Y have uniform distribution, thus :

$$E(X) = E(Y) = \frac{1}{2}$$

and $var(X) = var(Y) = \frac{1}{12}$. Now

$$\begin{aligned} E[XY] &= E[YE(X/Y)] \\ &= \frac{1}{2} \times \frac{1}{2}(1 - 2c) + \frac{1}{2}c \times c \times (\frac{1}{2} + c - c^2) \\ &+ c \times (\frac{1}{2} - c + c^2)(1 - \frac{1}{2}c) \\ &= \frac{1}{4} - c^2(1 - c)^2. \end{aligned} \tag{2.7}$$

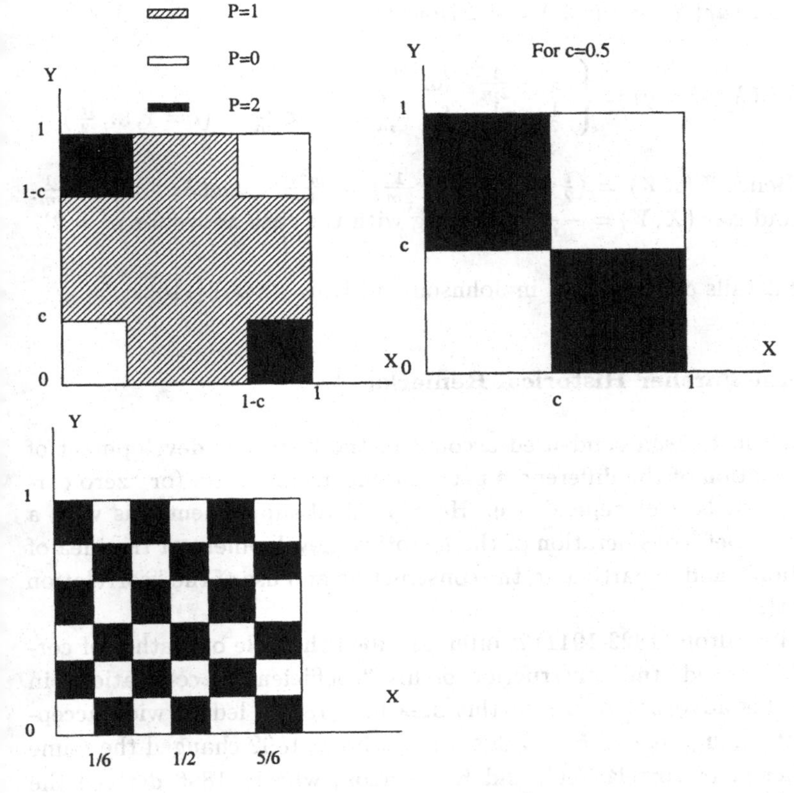

Fig. 2.3 *3a : Original pattern; 3b : with* $c = \frac{1}{2}$ *; 3c : Cherkerboard with m=6.*

Thus $cov(X, Y) = \frac{1}{4} - c^2(1-c)^2 - \frac{1}{4} = -c^2(1-c)^2$ and

$$corr(X, Y) = -12c^2(1-c)^2.$$

For $c = \frac{1}{2}$, $corr(X, Y) = -\frac{3}{4}$, and the pattern becomes the one presented in Figure 2.3-b :

(2) $m \times m$ checkerboard (m even) (Figure 2.3-c):
Both X and Y have uniform distributions. Thus

$$E(X) = E(Y) = \frac{1}{2}$$

and $var(X) = var(Y) = \frac{1}{12}$. However

$$E(X/Y = y) = \begin{cases} \frac{1}{2} + \frac{1}{2m} & \text{for } \frac{2(i-1)}{m} < y < \frac{2i-1}{m} \\ \frac{1}{2} - \frac{1}{2m} & \text{for } \frac{2i-1}{m} < y < \frac{2i}{m} \end{cases} \quad (i = 1, ..., \frac{m}{2}).$$

Hence $E(XY) = (\frac{1}{2} + \frac{1}{2m})(\frac{1}{2} - \frac{1}{2m}) = \frac{m^2-1}{4m^2}$, $cov(X, Y) = -\frac{1}{4m^2}$ and $corr(X, Y) = -\frac{3}{m^2}$ (agreeing with the case above for $m = 2$).

More details can be found in Johnson and Kotz (2000) [115].

2.4 Some Further Historical Remarks

Section 2.2 includes a condensed account of the historical development of the appreciation of the differences between uncorrelatedness (or "zero correlation") and lack of dependence. Here we shall supplement this with a broad, but brief, consideration of the historical development of the idea of "correlation", and in particular, the construction and use of the "correlation coefficient(s)" .

Francis Galton (1822-1911) is often ascribed the title of "father of correlation". Indeed, the construction of his "coefficient of co-relation" in 1885 and his advocacy of use of this measure (1888), led to wide acceptance of it, though it was F.Y. Edgeworth, who in 1892 changed the name to "coefficient of correlation", and K. Pearson, who in 1896 derived the analytic product-moment formula (often called the "Pearson correlation coefficient"). [Earlier this quantity had been called the "Galton" (or "Galton's") function.] This work of Galton has been described by his biographer Karl Pearson as "perhaps the most important of his writings". Recently several other scholars have corrected Pearson's story of the development of correlation so that a reasonably detailed account of the long intellectual gestation of this important tool is now available (see Fancher [68] for details). Fancher also describes a previously unpublished and overlooked study conducted by Galton in 1883 which deals with the relation between examination scores and later success studies in which he (Galton) essentially calculated the first correlation coefficient, albeit in a different standardized scale.

It thus took Galton five years to cast his correlated variables in what is now known probable errors terms. Indeed, important scientific breakthroughs have often followed long incubation periods of unsuccessful attempts and have appeared to be obvious only in retrospect.

Reaction to Galton's discovery in France was very swift. Two examples of immediate responses are illuminating. These were by Cheysson and Durkheim.

(1) Émile Cheysson (1836-1910), a disciple of Frédéric Le Play, was an engineer of bridges and a member of the Société Statistique de Paris. Quickly acquainted with Galton's lecture presented in September of 1885, he reviewed it shortly thereafter in an article published in *Le Temps* (the equivalent of today's *Le Monde*) on October 23, 1885.

(2) Similarly, Durkheim (1858-1917) in his famous book : "De la division du travail social" ("The division of Labor in Society" 1898) relies on Galton to show that the "social group" (rather than race) pulls the individual back towards its "average type".

However, there were predecessors to Galton. Notably : the American mathematician Robert Adrain (1775-1843), a native of Carrickfergus, Ireland, who taught mathematics at Rutgers University and the University of Pennsylvania; the renowned French mathematician Pierre-Simon Laplace (1749-1827); the Italian astronomer Giovanni Antonio Amadeus Plana (1781-1864); and the French physicist Auguste Bravais (1811-1863).

Each of these scholars approached the problem of measuring "dependence" from the point of view of a mathematical analyst, but did not appear to see (so far as we can discern) possible practical relevance of the formulas they constructed. It is quite likely that too close attention to the theoretical analysis, with limited experience of empirical data, contributed to this unfortunate situation.

In particular Laplace (1811), Plana (1818, 1820) and Gauss (1823) worked on what would now be called multinormal distributions (mostly bivariate, occasionally trivariate), but without special attention to the practical meaning of the coefficients of product terms in the exponent of

$$\exp(-Q(x_1, ..., x_n))$$

where $Q(.)$ is a quadratic function. Later, Bravais (1846), using a geometrical approach to derive the bivariate density function did mention the existence of a relationship ("correlation") being represented by the two variables, without giving the idea further discussion.

Forty years after Bravais in 1886, a British mathematician, J. Hamilton Dickson dealt with a special problem proposed to him by Francis Galton, and reached, on a somewhat narrow basis, some of Bravais's results for correlation of two variables. F. Galton at the same time introduced an improved notation, for the "Galton function" or coefficient of correlation. This does, indeed, appear in Bravais's work, but a single symbol is not used for it. As already has been mentioned above, in 1892 F.Y. Edgeworth, also unfamiliar with Bravais's memoir, dealt in a paper on "Correlated Averages" with correlation for three variables. He obtained results identical with Bravais, although expressed in terms of "Galton's functions".

C.R. Rao (1893), while reminiscing on the origin and development of the correlation coefficient, has pointed out that the source of symbol R for the correlation coefficient was (according to Karl Pearson) really the first letter of *Reversion.*

In his classical treatise on *Natural Inheritance* (1908), F. Galton wrote of his discovery, with the inspiring phraseology so typical of him : "This part of the inquiry may be said to run along a road on a high level, that affords wide views in unexpected directions, and from which easy descents may be made to totally different goals ..."

The product–moment (or the Pearson) correlation coefficient was first described explicitly by Karl Pearson (1896, p.265). He introduced the formula with the following words : "Thus it appears that the observed result is the most probable when r is given by the value $S(x,y)/n\sigma_x\sigma_y$. This value presents no practical difficulty in calculation, and therefore we shall adopt it. It is the value given by Bravais, but he does not show it is the best." The formula Pearson described is the one given in modern usage by :

$$r = \frac{\Sigma_{i=1}^{n} X_i^* Y_i^*}{n\sigma_X\sigma_Y},$$

where $X_i^* = X_i - \overline{X}$ and $Y_i^* = Y_i - \overline{Y}$. Numerous variants of this formula have been developed (see, e.g. Johnson *et al.* [116] for details).

H.P. Bowditch in his study "Growth of Children" in 1877 came close to the idea of correlation and not far from the idea of regression, but he was impeded by the fact that what he really was seeking was *partial* correlation.

2.5 A Brief Tour of Early Applications and Misinterpretations

In earlier applications of the correlation coefficient, measurements of two or more characters of the same individuals were frequently utilized for the measurement of correlation in biology and anthropology.

F. Galton has applied the method of measuring correlation to the number of Müllerian glands on the right and left sides of swine. The number of these glands varies widely with the individual. In case of absolute symmetry there would be the same number of glands on both right and left sides and the correlation would be perfect. As a matter of fact, although the correlation is not perfect, it is large. It is remarkable that such a specialized and narrow practical study would lead to the development of what is admittedly the most broadly applied index in all of statistics. The range of interpretations for the correlation coefficient shows the growth of this popular index over the past century. Rodgers and Nicewander (1988) [183] present 13 ways to look at the correlation coefficient. On the other hand, the index itself is surprisingly unchanged from the one originally proposed.

As we have already noted the basic idea was in the air at least 50 years before F. Galton. Rodgers and Nicewander (1988) present the table (Table 2.2) of landmarks in the history of correlation and regression.

F.Y. Edgeworth, explained in 1893 correlation as a "product-moment" (that is the normed total of the products of two correlated variables). But his formulation was not taken up and retranslated by others, and the standard expression of the coefficient of correlation is attributed to Pearson, who published it three years later. The difference between them is that Edgeworth (1893) entitled his article "Exercise in the Calculation of Errors" which was technical but established no link with anyone at all, whereas K. Pearson's title "Mathematical Contributions to the Theory of Evolution: Regression, Heredity, and Panmixia" managed to combine mathematics, Darwin, Galton and heredity (not to mention a new subject, that did not survive). Evolution and heredity were crucial points of interest for Pearson, and aroused the curiosity of a larger public than did Edgeworth's mathematical formulas.

By now, over a century later, contemporary scientists often take the correlation coefficient for granted. It is not appreciated that before Galton and Pearson, the only means to establish a relationship between variables was to deduce a *causative* connection. There was no way to discuss –let

Table 2.2 *Landmarks in the History of Correlation coefficient (based on Rodgers and Nicewander (1988))*

Date	Person	Event
1823	Carl Friedrich Gauss, German mathematician	Developed the normal surface of N correlate variates.
1843	John Stuart Mill, British philosopher	Proposed four canons of induction, includin concomitant variation.
1846	Auguste Bravais, French naval officer and astronomer	Referred to "une corrélation", studied bivari ate normal distribution.
1868	Charles Darwin, Galton's cousin, British natural philosopher	"All parts of the organisation are ... connecte and correlated."
1877	Francis Galton, British, the father of Biometrics	First discussed "reversion", the predecessor o regression.
1885	Francis Galton	Published bivariate scatterplot with norma isodensity lines, the first graph of correlation Completed the theory of bi-variate norma correlation.
1888	Francis Galton	Defined r conceptually, specified its uppe bound.
1893	F.Y. Edgeworth, British economist and theoretical statistician	Defined what is known as Pearson produc moment correlation coefficient but his discov ery remained unnoticed.
1895	Karl Pearson, British statistician	Redefined the Galton-Pearson product moment correlation coefficient.
1920	Karl Pearson	compiled "Notes on the History of Correla tion".

alone measure– the association between variables that lack a cause-effect relationship.

A.L. Bowley in his famous *Elements of Statistics* (first edition, pp. 316-319) and W.M. Persons in a paper in Publications of the American Statistical Assoc., Dec. 1910, pp. 298-299, characterizes the correlation coefficient as follows :

"The formula for the coefficient was found by assuming that a large number of independent causes operate upon each of the two series x and

y, producing normal distribution in both cases. Upon the assumption that the set of causes operating upon the series x is *not independent* of the set of causes operating upon the series y, the value $r = \frac{\Sigma xy}{n\sigma_x\sigma_y}$ is obtained. This value becomes zero when the operating causes are absolutely independent. "

A popular application of correlation coefficient in business and economic was introduced by W.M. Persons in his study and construction of the famous business barometer (Warren M. Persons : "Construction of a Business Barometer based upon Annual Data ", Amer. Econ. Review,Dec 1916, pp. 739-769). By computing the coefficient of correlation between cycles of relative wholesale prices and a series of statistics indicating business conditions when the prices series precedes(-) or lag behind (+) the others, Persons selects 9 series as a business barometer. See Table 2.3.

Table 2.3 *Coefficients of Correlation between Cycles of Relative Wholesale Prices and Cycles of Series Entering into the Business Barometer, 1879-1913 (Amer. Econ. Rev. Dec 1916 p. 757)*

Coefficient of Correlation Prices Lag behind (+) by :							
Series Correlated with Relative Wholesale Prices	- 2yr +	- 1yr +	0 yr +	+1 yr +	+2 yr +	+3 yr +	+4 yr +
Gross receipt of railroads	.847	.917	.945	.856	.748	.637	-
Net earnings of railroads	.690	.763	.862	.839	.803	.811	-
Coal produced	.787	.865	.931	.880	.795	.731	.630
Exports from the U.S.	.547	.671	.783	.786	.772	.328	-
Imports into the U.S.	.796	.796	.861	.754	.578	.445	-
Pig-iron produced	-	-	.756	.738	.631	.617	.528
Price of pig-iron	.406	.558	.763	.739	.637	.576	-
Immigration	.606	.718	.789	.626	.494	-	-
Relative wholesale prices	.811	.923	1.000	.923	.811	.691	.548

Other earlier applications are due to E. Davenport in the theory of breeding (E. Davenport :*The Principles of Breeding*, New-York, 1907), to E.L. Thorndinek and W. Brown in psychology (E.L. Thorndike : *Mental and Social Measurements*, New-York, 1913, William Brown :*The Essentials of Mental Measurement*, Cambridge (England), 1911) and somewhat later in zoology (J.A. Morris : *An Outline of Current Progress in the Theory of Correlation and Contingency*, Amer. Natur., Jan. 1916, Vol L, pp. 53-64).

The earlier statistical textbooks usually gave the frequently observed correspondence between the number of births and marriages and the price of grain as an illustration of correlated time-series. In a most interesting, but perhaps somewhat misleading contribution, R. H. Hooker (1901) has found that in England "the parallelism of marriages rates and foreign trade during the period 1861-1895 is not greatest for simultaneous fluctuations but for marriage rates and imports (or total foreign trade) which precede the marriages rates by a third of a year, or export which precede marriage rate by about half a year." This was interpreted that a certain amount of time must often elapse between the operation of cause and effect! (R. H. Hooker : *Correlation of the Marriage-Rate with Trade*, Journ. Royal Sta. Soc., 1901, p. 487)

It should be noted that L. March (1859-1933), Director of "Statistique Générale de la France", in the paper *Comparaison numérique de courbes statistiques*, Journ. Soc. Stat. de Paris, 1905, pp. 265-277 and 306-311, constructed a "coefficient de dépendance " based on elementary mathematical formulas, very similar to the Kendall's τ. By means of this coefficient, the correspondence of the fluctuations of two time series is summarized and an index is obtained that varies with the degree of coincidence of the series compared. More precisely, March's paper studies the dependence between the series of number of marriages per 10000 inhabitants and births during the period of 1873-1903 in France. He denotes, for each series of size n the change from year to year by + or - depending on whether the number increases or decreases and 0 if it remained steady, and counts the number of concordances and discordances between the two series. Next he calculates his index of matches (c) and mismatches (d) by $\frac{c-d}{n}$ or $\frac{c-d}{c+d}$. The second index is used when the 0's are absent. He notes that presumably these indices are due to G. Fechner in his famous *Kollectivmasslehre*, (Leipzig, 1897). Furthermore, March calculates the accuracy of the indices using Fechner's method for calculating the variance of the observed proportion of occurrence of certain events which he identified with $\frac{c-d}{c+d}$ to obtain the well known formula $0.67\sqrt{\frac{p(1-p)}{n}}$. To account for the amount of variation from year to year (not just its direction), March follows Cheysson's (1885) suggestion to standardize the Fechner-March index and arrives at the Galton-Pearson correlation coefficient (Their names appear in the discussion related to calculation of variation of the proposed coefficient). A minor historical curiosity : while all sources known to the authors of this book cite Bravais contribution to 1846, March refers to 1837 Bravais contribution.

The meaning (actually an interpretation) of the correlation coefficient is perhaps due to A.L. Bowley. This interpretation was for many years dominant and accepted as an undisputable doctrine, almost carved in stone.

"When r is not greater than its probable error $(0.67\frac{1-r^2}{\sqrt{n}})$ we have no evidence that there is any correlation, for the observed phenomena might easily arise from totally unconnected causes; but, when r is greater than, say, 6 times its probable error, we may be practically certain that the phenomena are not independent of each other, for the chance that the observed results would be obtained from unconnected causes is practically zero."

Studying G.U. Yule's *"Theory of Correlation"* (Journ. Roy. Stat. Soc., Vol LX (1897), p.813) is recommanded for anyone interested in the problems of the origin, the evergreen meaning and applications of this widely used (and abused) concept. It should always be kept in mind that the theory of correlation has been constructed on the basis of the two-dimensional normal distribution. However transformation to normality is not always the best answer to the limitations of correlation coefficients. Depending on the context such transforms can distort and obscure important relations, especially if certain subranges of the distribution are of greater scientific or clinical importance than others (see S. Greenland (1996) for a detailed discussion).

It ought to be noted that the prevalent interpretation of the theory of correlation is as a theory of the concomitant variation of two or more attributes of a group of individual entities, the attributes being measured with respect to each entity, while the Galton-Pearson coefficient is just a numerical measure of the linear concomitant variation. The intricate relation between causality and correlation attracted attention of philosophers in the second half of the 20th century. For an excellent survey, see G. Irzik

[105] (1996).

So far our interest was centered around the properties of bivariate distributions with zero correlation. It is also instructive to point out that correlation near one in magnitude does not imply that a relationship is near perfect linearity. To clarify this point we present a construction due to Turner (1997), which is somewhat similar to our notched and nicked modifications discussed above.

Let discrete random vector has joint uniform distribution on the 4 points $(-k, -a-\epsilon)$, $(-k, -a+\epsilon)$, $(k, a-\epsilon)$, $(k, a+\epsilon)$. See Figure 2.4 for the case of k, a, ϵ, and $a-\epsilon$ all being positive.

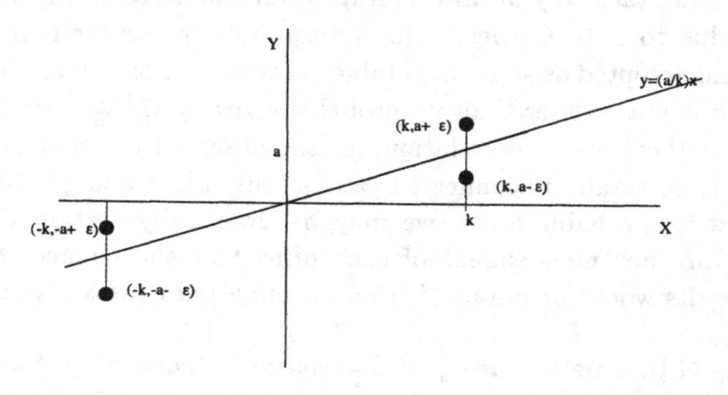

Fig. 2.4 : *Uniform distribution at 4 points $(\pm k, a \pm \epsilon)$, when a, k, ϵ, and $a-\epsilon$ are positive. The regression line $y=ax/k$ is also plotted.*

For these variables $E(X) = E(Y) = 0$, $var(X) = k^2$ and $var(Y) = a^2 + \epsilon^2$ and $cov(X,Y) = ak$. Thus the correlation coefficient (for a positive k) is :

$$corr(X,Y) = \frac{a}{(a^2 + \epsilon^2)^{1/2}} \, .$$

For any $c \in (-1, 1)$, the parameters a and ϵ can be manipulated to have $corr(X,Y) = c$. Given c and ϵ and solving for $corr(X,Y) = c$, we have $|a| = \frac{|c\epsilon|}{(1-c^2)^{1/2}}$. Note that as $|a| \to \infty$, $|corr(X,Y)| \to +1$ for any fixed $\epsilon \neq 0$, but the points $(k, a+\epsilon)$ and $(k, a-\epsilon)$ remain $2|\epsilon|$ units apart (as well as the other two points) and the perfect linearity is not achieved no matter how close $|corr(X,Y)|$ is to 1. Also for the case $a = k$ with $k > 0$, $|corr(X,Y)| = \frac{k}{(k^2+\epsilon^2)^{1/2}} \to 1$, *as* $k \to \infty$, but the points are bounded away from the regression line in all directions.

Chapter 3
Concepts of Dependence and Stochastic Ordering

3.1 Introduction

Dependence relations between random variables is one of the most widely studied topics in probability theory and statistics. Unless specific assumptions are made about the dependence no meaningful statistical model can be constructed. As we shall see below the concept of covariance plays an important role in defining dependence between variables. A term "statistical dependence" rather than dependence should perhaps be used to emphasize that we are dealing with new concepts specific for probability theory which may not coincide with the daily meaning of this concept.

K Pearson (1857-1956) seems to be one of the originators of the dependence concepts and measures in statistical methodology. C. Gini (1884-1966) [86] and M. Fréchet (1878-1913) [74] have also contributed substantially to the theory in its early stages.

Numerous concepts of positive dependence are delineated in the literature. Pioneering sources, in the second half of the 20th century are the paper of Harris [92] (1960) and Lehmann [142] (1966), followed by Esary, Proschan and Walkup [64] [65], Kimeldorf and Sampson [129] and many others. Harris [93] and Shaked [201] define additional concepts appropriate for survival variables, and show how they are nested. Joe [108] characterized the distributions for which dependence is concentrated at the lower and upper tails. These concepts initially defined for two variables has been extended to a multivariate random vector $(X_1, X_2, ..., X_n)$ with $n \geq 2$.

When $n = 2$ the negative dependence is easily constructed by reversing the concepts of positive dependence, this is carried out in Lehmann's paper.

However for $n > 2$ negative dependence is not a simple mirror of positive dependence. Using the paradigm of the multinomial distribution, Block *et al.* [22] have developed conditions to obtain sub-families of negatively dependent distributions. Joag-Dev and Proschan [106] have defined negative association using the idea that if a set of negatively random variables is split into two subsets, then when one subset tends to be "large", the other tends to be "small".

Bivariate dependence can be generalized in another way by considering that the random vector $(X_1, X_2, ..., X_n)$ can be split into k sets and we may be more interested in the relationships *between* the k sets than the relationships *within* each set. This situation is encountered for example in complex engineering systems where the relationships among subsystems and the relationships within subsystems may need to be studied separately. Chehhtry, Kimeldorf and Sampson [43] have developed these notions of setwise dependence.

Generalizing a bivariate concept of positive dependence leads to a definition of an ordering which compares two bivariate distributions to determine whether one distribution is more concentrated than the other. Kimeldorf and Sampson [129] [130] have provided a general framework for studying positive dependence orderings. Yanagimoto and Okamoto [226] [228], Schriever [193], Shaked and Tong [203], Rinott and Pollack [182], Capéraà and Genest [37], Fang and Joe [69], Chang [41], Shaked and Shantikumar [205], Scarsini [191], Bäuerle [18] and Müller and Scarsini [158], [159] have studied specific orderings.

Starting from two partial orderings, Yanagimoto [228], Metry and Sampson [155] have generated systematically all the intermediate partial orderings.

Using the Bayesian paradigm, Brady and Singpurwalla [31] have defined a stochastic dependence and a stochastic ordering between two variables (and two distributions).

Section two presents concepts of bivariate positive dependence and their generalizations for more than two variables. In the third section we present a few results concerning negative dependence in the case of more than two variables. In the fourth we extend these notions to the setwise dependence. The fifth section enumerates a few properties and applications which are not presented in that book. The sixth section transfers the notion of bivariate positive dependence to positive dependence orderings. The seventh provides additional information concerning stochastic dependence and ordering.

3.2 Concepts of Positive Dependence

A motivation of Lehmann [142] when he defined the basic concepts of positive dependence was to provide tests of independence between two variables that are not biased. Actually, here, to construct an unbiased test we ought to stipulate the alternative hypothesis. Lehmann identified sub-families of bivariate distributions for which this property of unbiasedness is valid. Later on, Kimeldorf and Sampson [129] exhibited properties that a sub-family of distributions $\mathcal{F}^+$ with given marginals has to obey to have a positive dependence property.

We present a few concepts for the case of two variables, before generalizing to the case of more than two variables. We denote by $F_1(x)$ and $F_2(y)$, the two marginals of $F(x,y)$. Recall that each distribution $F(x,y)$ has upper and a lower Fréchet bounds, which are $F^+(x,y) = min(F_1(x), F_2(y))$ and $F^-(x,y) = max(0, F_1(x) + F_2(y) - 1)$ respectively. Depending on the context, we write $(X,Y) \in \mathcal{F}^+$ or $F \in \mathcal{F}^+$.

3.2.1 *The Kimeldorf and Sampson conditions*

The sub-family $\mathcal{F}^+$ possesses the following properties :

(1)

$$F \in \mathcal{F}^+ \Rightarrow F(x,y) \geq F_1(x)F_2(y), \quad \forall x \; \forall y.$$

This condition is the positive quadrant dependence (PQD), which is studied in the next section.

(2) If $F(x,y) \in \mathcal{F}^+$, so does $F^+(x,y)$.
(3) If $F(x,y) \in \mathcal{F}^+$, so does $F^0(x,y) = F_1(x)F_2(y)$.
(4) If $(X,Y) \in \mathcal{F}^+$, so does $(\phi(X),Y) \in \mathcal{F}^+$, where Φ is an increasing function.
(5) If $(X,Y) \in \mathcal{F}^+$, so does (Y,X).
(6) If $(X,Y) \in \mathcal{F}^+$, so does $(-X,-Y)$.
(7) if $F_n \in \mathcal{F}^+$ and $F_n \xrightarrow{L} F$, then $F \in \mathcal{F}^+$. (where the symbol "L" means "converges in distribution").

3.2.2 The positive quadrant dependence (PQD)

It is a rather weak condition of positive dependence :

$$\forall x, \forall y, F(x,y) \geq F_1(x)F_2(y). \tag{3.1}$$

It is equivalent to

$$\forall x, \forall y, S(x,y) \geq S_1(x)S_2(y). \tag{3.2}$$

If we reverse the inequalities 3-1 or 3-2, we have negative quadrant dependence (NQD).

The PQD-family satisfies the seven conditions of Kimeldorf and Sampson.

The condition of PQD implies that $cov(X,Y) \geq 0$. Namely, using a fundamental lemma due to Hoeffding (1940) in an obscure publication and reprinted in his collected works [96], we know that :

$$cov(X,Y) = \int_{-\infty}^{+\infty}\int_{-\infty}^{+\infty} (F(x,y) - F_1(x)F_2(y))dxdy. \tag{3.3}$$

The proof is given in Lehmann in [142] and we sketch here it with minor modifications. Let (X_1, Y_1) and (X_2, Y_2) be independent and distributed according to $F(x,y)$. Then

$$\begin{aligned} &2\left(E(X_1Y_1 - E(X_1)E(Y_1)\right) \\ &\quad = E(X_1 - X_2)(Y_1 - Y_2) \\ &\quad = E\int\int (I_{]-\infty,X_1]}(u) - I_{]-\infty,X_2]}(u))(I_{]-\infty,Y_1]}(v) - I_{]-\infty,Y_2]}(v))\, dudv. \end{aligned}$$

Since $|E(X,Y)|$, $E|X|$ and $E|Y|$ are assumed to be finite, we can exchange the expectation and the integral signs, and subsequent calculations result is twice the right hand side of (3.3). This completes the proof.

This implies that in the family of PQD distributions if the covariance is zero, then the variables are independent.

Conditions (3.1) or (3.2) are equivalent to (Esary *et al.* [64]) :

$$cov(f(X), g(Y)) \geq 0, \qquad \forall f,\ \forall\, g \text{ increasing functions}$$

To show that this condition implies PQD, Esary uses two indicators functions $f(X) = I_{]-\infty,x]}(X)$ and $g(Y) = I_{]-\infty,y]}(Y)$, which are increasing and verify $E(f(X)) = P(X \leq x)$ and $E(g(Y)) = P(Y \leq y)$. The reverse statement is a lemma due to Lehmann [142].

Property : The Hoeffding lemma can be extended, using the supermodular functions (see Cambanis (1976) [36], Tchen (1980) [216], Quesada-Molina (1992) [174]) : Let (X^*, Y^*) be an independent random vector with X^* and Y^* having the same distribution as X and Y respectively, and $\phi : R^2 \mapsto R$ a supermodular right-continuous function, such that the expectations of $E(\phi(X,Y))$ and $E(\phi(X^*,Y^*))$ exist, then

$$E(\phi(X,Y)) - E(\phi(X^*,Y^*)) = \int_{-\infty}^{+\infty}\int_{-\infty}^{+\infty} (F(x,y) - F_1(x)F_2(y))d\phi(x,y).$$

The proof use similar arguments as the Hoeffding lemma.

Corollary : A consequence of this is that the property that (X,Y) PQD-dependent $\Rightarrow E(\phi(X,Y)) \geq E(\phi(X^*,Y^*))$ for all the supermodular functions $\phi : R^2 \mapsto R$. The reverse is also true using the fact that the indicator functions are supermodular and right-continuous.

3.2.2.1 *Positive upper or lower orthant dependence*

Let $\mathbf{x} = (x_1, x_2, ..., x_n)$ and $\mathbf{y} = (y_1, y_2, ..., y_n)$ be two vectors in R^n. We say that $\mathbf{x} > \mathbf{y}$ if for all $i = 1, 2, ..., n$ $x_i > y_i$.

Let $\mathbf{X} = (X_1, X_2, ..., X_n)$ $(n > 2)$ be a vector of random variables. If for every $\mathbf{x} = (x_1, x_2, ..., x_n)$

$$P(\mathbf{X} > \mathbf{x}) \geq \prod_{i=1}^{n} P(X_i > x_i) \tag{3.4}$$

we then say that $\mathbf{X} = (X_1, X_2, ..., X_n)$ is positively upper orthant dependent (PUOD). If, for every $\mathbf{x}$, we have :

$$P(\mathbf{X} \leq \mathbf{x}) \geq \prod_{i=1}^{n} P(X_i \leq x_i) \tag{3.5}$$

we then say that $\mathbf{X} = (X_1, X_2, ..., X_n)$ is positively lower orthant dependent (PLOD). The two cases (3.4) and (3.5) are equivalent only for $n = 2$.

Proposition : Connection with supermodular functions.

Let $\mathbf{X}$ be a random vector with marginals $F_i(x_i)$, $i = 1, ..., n$. If for all supermodular function $\phi : R^n \mapsto R$, $\mathbf{X}$ satisfies

$$E(\phi(\mathbf{X})) > E(\phi(\mathbf{X}^*))$$

where $\mathbf{X}^*$ is an n-random vector having the distribution $\Pi_{i=1}^n F_i(x_i)$, then the vector $\mathbf{X}$ is positively upper-orthant and lower-orthant dependent. To prove this, it suffices to consider the two indicator functions $\phi_{\mathbf{x}}(\mathbf{z}) = \mathbf{I}_{\mathbf{z}>\mathbf{x}}(\mathbf{z})$ and $\psi_{\mathbf{x}}(\mathbf{z}) = \mathbf{I}_{\mathbf{z}\leq\mathbf{x}}(\mathbf{z})$ which are supermodular (see for example Shaked and Shantikumar [204]).

3.2.2.2 *Positive upper or lower set dependence*

For this definition we require a notion which is more general than "orthant". A set U in R^n is called an upper set if $\mathbf{x} \in U$ and $\mathbf{y} > \mathbf{x}$ (i.e. for all i $i = 1, ..., n$, $y_i > x_i$) implies $\mathbf{Y} \in U$. A lower set L is the complement of an upper set U (however lower orthant is not the complement of upper orthant!).

A definition stronger than positive upper orthant dependence is positive upper set dependence given by :

$$P(\mathbf{X} \in \bigcap_k U_k) \geq \prod_k P(\mathbf{X} \in U_k) \tag{3.6}$$

where U_k, $(k \geq 2)$ are upper sets in R^n.

3.2.3 *Association*

Esary *et al.* [65] have defined (positive) association between X and Y, by the property :

$$Cov(f(X,Y), g(X,Y)) \geq 0 \; \forall \, f, \; \forall \, g \text{ increasing functions,}$$

provided the covariances exist. This property implies PQD. The family of all associated distributions also satisfies the conditions of Kimeldorf and Sampson.

In the same manner, we can define positive association for a random vector $\mathbf{X} = (X_1, X_2, ..., X_n)$ by the relation :

$$cov(g_1(\mathbf{X}), g_2(\mathbf{X})) \geq 0$$

whenever g_1 and g_2 are two real valued increasing functions.

We enumerate a few immediate properties of positive association.

(1) Any subset of associated random variables are associated.
(2) If two subsets of associated random variables are independent, then their union is associated.
(3) A single random variable is associated.
(4) Increasing functions of associated random variables are associated.

A consequence of property 2 is that independent random variables are associated. Esary *et al.* [64] also show that for binary random variables, association is equivalent to $cov(X, Y) \geq 0$. Another property is that in the definition of association, it is sufficient to consider *bounded continuous* functions f and g. It is also sufficient to consider indicator functions.

Moreover, Esary *et al.* [64] have shown that positive association is equivalent to positive upper (open) set dependence, and therefore, in particular, positive association implies PUOD.

A simple consequence is that the larger dependence in $\mathbf{X}$ is associated with the larger dispersion of $T = \Sigma_{i=1}^{n} X_i$. Namely, $var(T) = nvar(X_1) + n(n-1)cov(X_1, X_2)$ (the one-dimensional marginals being identical). Therefore positive dependence in $\mathbf{X}$ produces overdispersion in $\mathbf{T}$ [228].

3.2.4 *Positive function dependence*

Let exchangeable random variables (X, Y) have symmetric joint distribution F. Shaked [202] has defined F to have positive function dependence if F is a positive defined kernel on the support $S \times S$ of (X, Y). He has discovered a characteristic property which is :

$$Cov(h(X), h(Y)) \geq 0$$

for every *real* function h, such that the covariance exists.

A multivariate extension can then be defined as :

$$E[\Pi_{i=1}^{n} h(X_i)] \geq \Pi_{i=1}^{n} E[h(X_i]$$

for all real valued h, such that the expectations exist.

3.2.5 *Positive regression dependence (PRD)*

Inequalities 3.1 and 3.2 can be rewritten as :

$$\forall x,\ \forall y,\ \frac{F(x,y)}{F_1(x)} \geq F_2(y)$$

i.e.

$$\forall x,\ \forall y,\ F(y/X \leq x) \geq F_2(y)$$

and similarly,

$$\forall x,\ \forall y,\ S(y/X \geq x) \geq S_2(y)$$

respectively.

A stronger condition would be :

$$\forall y, F(y/X \leq x) \text{ is non-increasing in } x$$

and,

$$\forall y, S(y/X \geq x) \text{ is non-decreasing in } x.$$

Esary *et al.* [64] designate that the distribution is, left tail decreasing (LTD) if the first case is valid and right tail increasing (RTI) if the second holds. It is easy to prove that in the bivariate case, these two conditions are equivalent.

A stronger condition than the preceding one is :

$$\forall y,\ S_{Y/x}(y) = S(y/X = x) \tag{3.7}$$

is an increasing function in x, or equivalently :

$$\forall y,\ F_{Y/x}(y) = F(y/X = x) \tag{3.8}$$

is a decreasing function in x. Lehmann calls this condition Positive Regression Dependence (PRD), other authors use the term Stochastically Increasing (SI).

Proposition(Capéraà and Genest, 1990) : If the conditional distribution $F_{Y/x}(y)$ is continuous, and strictly increasing, then it has an inverse $F^{-1}_{Y/x}(u)$, and we can define, without ambiguity, a cumulative distribution function which maps $[0,1]$ to $[0,1]$, $F_{x',x}(u) = F_{Y/x'} \circ F_{Y/x}^{-1}(u)$. The PRD property is then equivalent to :

$$\forall x < x', \; \forall\, 0 \leq u \leq 1, \;\; F_{x',x}(u) \leq u.$$

3.2.5.1 *Multivariate case*

There are numerous multivariate generalizations of PRD. We shall cite three of them. The third, proposed by Müller and Scarsini [159], seems to be more natural and is stronger than the second.

Definition 1 PDS : The random vector $\mathbf{X} = (X_1, X_2, ..., X_n)$ $n \geq 2$, is positive dependent through the stochastic ordering (PDS) if $P(X_i > x_i/X_j = x)$, is increasing in x for all $1 \leq j \leq n$, $j \neq i$ and all x_i.

Definition 2 CIS : The random vector $\mathbf{X} = (X_1, X_2, ..., X_n)$ $n \geq 2$, is conditionally increasing in sequence (CIS) if $P(X_i > x_i/X_j = x_j,\ j = 1, ..., i-1)$ is increasing in $x_1, x_2 ..., x_{i-1}$ for all x_i, $i = 1, ..., n$.

Definition 3 CI : The random vector $\mathbf{X} = (X_1, X_2, ..., X_n)$ $n \geq 2$, is conditionally increasing (CI) if $P(X_i > x_i/X_j = x_j,\ j \in J)$ with $J \subset \{1, 2..n\}$ is increasing in x_j for all x_j, $j \in J$ and for all x_i. One can see that $\mathbf{X}$ is CI if and only if $\mathbf{X}_\pi = (X_{\pi(1)}, X_{\pi(2)}, ..., X_{\pi(n)})$ is CIS for all permutations π of $\{1, 2, ..., n\}$.

Barlow and Proschan [16] show that the CIS property implies positive association.

3.2.6 *The Lihelihood ratio dependence (LRD)*

The LRD dependence is a property of total positivity of order two (TP_2) for the density $f(x,y)$. The ideas of total positivity are developed in Karlin [123], see Chapter 1. Specifically $f(x,y)$ is TP_2 if :

$$\begin{cases} f(x,y) \geq 0 \\ \forall x < x', \forall y < y', \;\; f(x,y)f(x',y') \geq f(x,y')f(x',y). \end{cases}$$

If this condition is fulfilled, then PRD and PQD properties are also fulfilled.

Using the distribution function $F_{x'x}(u) = F_{Y/x'} \circ F_{Y/x}^{-1}(u)$, defined in Section 3.2.5, under the same hypotheses ($F_{Y/x}(y)$ continuous and strictly increasing for all x), it is easy to prove that the LRD-property is equivalent to :

$$\forall x < x' \quad , \forall 0 < u < t < v < 1 \quad \frac{F_{x'x}(t) - F_{x'x}(u)}{F_{x'x}(v) - F_{x'x}(u)} \leq \frac{t-u}{v-u} \qquad (3.9)$$

The last relation is equivalent to the convexity of $F_{x'x}(u)$ on $[0, 1]$.

There are numerous generalizations of TP_2. One is TP_2 in pairs, if for any pair of arguments x_i, x_j the density $f(x_1, ..., x_i, ..., x_j, ...x_n)$ viewed as a function of x_i, x_j with the remaining arguments fixed is TP_2. Barlow and Proschan [16] have shown that this last property implies PLOD and PUOD.

The definition of TP_2 in pairs is equivalent in R^n to the MTP_2 property (Multivariate Totally Positive of order two, MTP_2) studied by Karlin and Rinott [124]. Namely:

$$f(\mathbf{x} \vee \mathbf{y}) f(\mathbf{x} \wedge \mathbf{y}) \geq f(\mathbf{x}) f(\mathbf{y}) \ , \forall \, \mathbf{x} \, \forall \, \mathbf{y} \in R^n$$

where $\mathbf{x} \vee \mathbf{y}$ denotes the vector of the maxima $max(x_i, y_i)$, $i = 1, ..., n$ and $\mathbf{x} \wedge \mathbf{y}$ the vector of the minima. The equivalence has been proved by Kemperman [125]. The property of MTP_2 is inherited by all the marginals densities of order two and higher.

Müller and Scarsini [159] show that the MTP_2 property implies CI. Namely MTP_2 implies CIS (as shown by Karlin and Rinott [124]), but the MTP_2 property is invariant under permutations. Thus if $\mathbf{X}$ is MTP_2, then $\mathbf{X}_\pi$ is CIS for all permutations π, and hence $\mathbf{X}$ is CI. The implications $MTP_2 \rightarrow CI \rightarrow CIS$ are strict. However in case of multinormal distributions MTP_2 and CI are equivalent if the covariance matrix is invertible [159].

3.2.7 Dependences DTP(m,n)

Using the notion of TP_2, Shaked [201] proposes some nested definitions of dependence, adapted for survival variables :

-the strongest notion of dependence is LRD, which Shaked names DT-P(0,0), when the density f is TP_2.

- Next he defines DTP(0,1) (resp DTP(1,0)), when $-D_1S$ (resp $-D_2S$) is TP_2.

- DTP(1,1), when S is TP_2.

- DTP(1,2), when the expected conditional residual life $m(Y/X > x) = E(Y - y/Y > y, X > x)$ is TP_2.

- DTP(m,n), when

$$\int_y^\infty \int_{y_{n-1}}^\infty \cdots \int_{y_1}^\infty \int_{x_{m-1}}^\infty \cdot\cdot \int_x^\infty \cdot \int_{x_1}^\infty f(x_0, y_0) dx_0 ... dx_{m-1} ... dy_0 ... dy_{n-1}$$

$\forall m > 1, \forall n > 2$ is TP_2.

Each notion of dependence implies the next one, and DTP(0,1) implies the notion of Lehmann's PRD.

From the TP_2 property of $-D_2S$ (resp $-D_1S$) one can easily deduce that the DTP(0,1) dependence (resp DTP(1,0)) is equivalent to the decrease w.r.t to x of the conditional hazard $h_{2/1}(x,y) = \frac{f(y/X=x)}{S(y/X=x)}$ (resp $h_{1/2}(x,y)$). See Chapter 1 for the connection between these two conditional hazards.

From the TP_2 property of S, one deduces that the DTP(1,1) dependence is equivalent to the decrease w.r.t. x of the conditional hazard $h_2(x,y)$, and also the decrease w.r.t. y of the conditional hazard $h_1(x,y)$. This concept of dependence is therefore the same as the concept of Right Corner Set Increasing (RCSI) of Harris. [93]

- the DTP(1,2) dependence is equivalent to the increase w.r.t. x of the expected residual life $m(y/X > x)$.

Denoting $S_{xx'}(u)$ the survival function associated to the c.d.f $F_{xx'}(u)$ defined Section 3-2-6, we have the following result : The DTP(0,1) definition is equivalent to :

$$\forall x < x' \quad \forall 0 < u < v < 1 \quad \frac{S_{x'x}(u)}{S_{x'x}(v)} \leq \frac{1-u}{1-v}. \tag{3.10}$$

Compare with relation (3.9).

3.2.8 *Positive dependence by mixture*

In various contexts, specifically in reliability theory and genetic studies, positive dependence by mixture is often assumed.

If (X,Y) are two random variables, independent conditionally with respect to a (latent) variable W with distribution function G, then their joint distribution function is :

$$F(x,y) = \int F^w(x) F^w(y) dG(w) ,$$

where $F^w(x)$ (resp $F^w(y)$) is the distribution function of X (resp Y) given W. In that case, one says that (X,Y) are positive dependent by mixture. Using the properties of TP_2 functions it is easy to associate a concept of dependence to the pair (X, Y). Namely if the distribution functions of the pair (X, W) and (Y, W) are DTP(m,0) and DTP(n,0) respectively, then the pair (X, Y) is DTP(m,n) (see, for example Shaked (1977) [201]). In particular (X, Y) is DTP(0,0) (i.e. have LRD dependence) if (X, W) and (Y, W) have LRD dependence. See Chapter 4 Sections 4.5.6 and 4.6 for applications.

A multivariate generalization is straightforward :

$$F(\mathbf{x}) = \int \Pi_{i=1}^{n} F^{w}(x_i) dG(w)$$

where $\mathbf{X}$ is a random exchangeable vector with independent components X_i given $W = w$.

A particular case is when $X_i = U_i + W$, $i = 1, ..., n$, with the $U_1, ..., U_i, ..., U$ independent. If U_i has a density f_i which is a Pólya function of order 2 (i.e. $f_i(u_i - w)$ is TP2), then each pair (X_i, X_j) is TP_2.

Examples :

If U_1, U_2, W are three independent variables, from the exponential distribution $\Gamma(1, 1)$, then $X = U_1 + W$ and $Y = U_2 + W$ are dependent, with the same distribution $\Gamma(2, 1)$. The pair (X, Y) with the density $f(x, y) = e^{-max(x,y)}(1 - e^{-min(x,y)})$ when x and y are positive and zero elsewhere is LRD, since the exponential density is Pólya of order 2 [123], and therefore also PRD and PQD.

Yanagimoto [228] provides the following simple example: If $U_i \sim N(0, \sigma^2)$, $W \sim N(0, \delta^2)$, then $\mathbf{X} \sim N(\mathbf{0}, \sigma^2 I + \delta^2 ee')$ (I being the identity matrix and the vector $e = (1, ..., 1)'$. The dependence between the pairs (X_i, X_j) is therefore increasing in δ^2.

3.2.9 *Implications of the concepts*

We have the following implications of the concepts described above:

For $n = 2$ variables:

$$\begin{array}{llll} LRD \rightarrow & DTP(0,1) \rightarrow & DTP(1,1) \rightarrow ... \rightarrow DTP(n,m) & \\ & \searrow PRD \rightarrow & RTI \text{ and } LTD \rightarrow Association \rightarrow PQD & \end{array}$$

For $n > 2$:

$$MTP_2 \rightarrow \quad CI \rightarrow CIS \rightarrow \quad Association \rightarrow PUOD \text{ and } PLOD$$

$$\text{and } PDS \rightarrow PUOD \text{ and } PLOD$$

3.2.10 *Lower and upper tail dependence*

In the case of copulas (see Chapter 4), where (U, V) is a pair of uniform variables on the unit square, Joe [108] defines upper tail dependence by :

$$\delta = lim_{u \to 1-} \frac{\bar{C}(u,u)}{1-u} > 0$$

is not zero. We also have :

$$\delta = lim_{u \to 1-} Pr(U > u / V > u) .$$

Similarly, the lower tail dependence holds if $\frac{C(u,u)}{u}$ has a limit γ different from zero, when u tends to zero. This limit is also :

$$\gamma = lim_{u \to 0+} Pr(U \leq u / V \leq u) .$$

There is a connection between the tail dependence of C and of the associated copula $C'(u,v) = \bar{C}(1-u, 1-v)$: the lower tail dependence of C is the upper tail dependence of C' and vice-versa. Indeed :

$$lim_{u \to 1-} \frac{\bar{C}(u,u)}{1-u} = lim_{u \to 0} \frac{\bar{C}(1-u, 1-u)}{u} \quad (3.11)$$

$$= lim_{u \to 0} \frac{C'(u,u)}{u} \quad (3.12)$$

These concepts of tail dependence are useful in study of dependence in extreme value distributions. See for example Joe [111], Tawn [215] and Currie [51].

3.3 Negative Dependence for More than Two Variables

Having defined several concepts of positive dependence for the case of two variables, we can easily obtain the analogous concepts of negative dependence. Namely if (X, Y) has a positive dependence, then $(X, -Y)$ on R^2, or if we have a constraint of positivity $(X, 1-Y)$ (on the unit square), have a negative dependence. However if we have more than two variables reversing definitions of positive dependence concepts do not allow us to retain the same appealing properties, in particular closure under marginalization, and nesting of the concepts according to their dominance (Section 3.2.9) do not exist here. The reason is that the perfect negative dependence does not exist for more than two variables (for three variables, for example, if one variable X_1 has a perfect negative dependence with X_2 and with X_3, then X_2 and X_3 have perfect positive dependence) and the larger n is the weaker is the negative dependence. (As n tends to infinity, the correlation between two variables from a set of n exchangeable variables tends to zero, see Chapter 6). Moreover some concepts of positive dependence may coexist with negative dependence (see for example Section 3.3.4).

We present some concepts to define negative dependence in the case of more than two variables.

3.3.1 *NUOD and NLOD*

If we reverse the inequalities (3.4) and (3.5), then we obtain the concepts of Negative Upper Orthant Dependence (NUOD), and Negative Lower Orthant Dependence (NLOD). A problem which has occupied many statisticians has been to exhibit conditions under which these inequalities are fulfilled. Except for the bivariate case, the conditions are more difficult to show than in the case of positive dependence.

Remark : We note that if (X_1, X_2, X_3) are NUOD, then $Y = X_1 + X_2$ and X_3 are not necessarily NUOD.

3.3.2 *Definition from RR_2*

In Section 3.2.7, we have defined the TP_2 property for a density. If we reverse the inequality in it then we have a property of negative dependence, which Karlin [123] refers to as RR_2, i.e. Reverse Regular of Order Two. However sufficient condition obtained by Barlow and Proschan [16] in the case of positive dependence (density TP2 in pairs implies PUOD) cannot be reversed : RR_2 in pairs for the density does not imply NUOD. Even if we impose additionally that the marginal densities are also RR_2 in pairs we have no guarantee that NUOD or NLOD is fulfilled.

We are thus required to find a stronger property than RR_2 in pairs.

3.3.3 *Structural condition*

The idea of Block *et al.* [22] is to search for a condition which mimics the dependence of the variables having a multinomial distribution, i.e. those for which the sum of the variables is fixed.

Definition : The random vector $(X_1, X_2, ..., X_n)$ satisfies the structural condition if there exist $n+1$ independent random variables $S_0, S_1, ..., S_n$, whose densities are Pólya of order two PF2 ($f(x-y)$ is TP_2 on $R \times R$) and a real number s, such that:

$$(X_1, X_2, ..., X_n) \stackrel{L}{=} [(S_1, S_2..S_n)/S_0 + S_1 + ... + S_n = s] .$$

Theorem ([22]) : if $(X_1, X_2, ..., X_n)$ satisfies the structural condition, then it is RR_2 in pairs and also NUOD and NLOD.

In particular, let $X_1, X_2, ..., X_k$ be k independent random variables each having PF_2 density, then the joint conditional distribution of $X_1, X_2, ..., X_k$ given ΣX_i is RR_2 in pairs.

Examples :

(1) Multinomial Distribution :
The multinomial distribution is the conditional distribution of independent Poisson random variables given their sum. Thus, by the theorem, it is RR_2 in pairs and also NUOD and NLOD.

(2) Multivariate Hypergeometric Distribution :
The multivariate hypergeometric distribution is the conditional distribution of independent binomial random variables given their sums. Thus, by the theorem, it is RR_2 in pairs and also NUOD and NLOD.

(3) Dirichlet Distribution :
The Dirichlet distribution is the conditional distribution of independent gamma random variables given their sum equals to one. Thus, by the theorem, it is RR_2 in pairs, and NUOD and NLOD.

(4) Multinormal Distribution:
Let $\mathbf{X} = (X_1, ..., X_n)$, be a multivariate symmetric normal random vector, with $corr(X_i, X_j) = \rho \leq 0$, $1 \leq i < j \leq n$. Then $\mathbf{X}$ is RR_2 in pairs. To see this, one supposes, without loss of generality that $E(X_i) = 0$ and $var(X_i) = 1$. And one considers the random variables $Y_1, ..., Y_n$ with independent identically distributed normal distribution such that $E(Y_i) = 0$, $var(Y_i) = 1 - \rho$, $i = 1, ..., n$ and an independent normal random variable Y_0 with $E(Y_0) = 0$ and $var(Y_0) = -\frac{(1-\rho)(1+(n-1)\rho)}{\rho}$. Then $(X_1, ..., X_n) \stackrel{L}{=} [(Y_1, ..., Y_n)/Y_0 + Y_1 + ... + Y_n = 0]$. Thus, $\mathbf{X}$ is RR_2 in pairs.
Application [228] : Let $U_i \sim N(0, \sigma^2)$, and $U_0 \sim N(0, \tau^2)$, then $\mathbf{X} \sim (U_1, ..., U_i, ..., U_n)/(\Sigma_{i=1}^n U_i + U_0)$ has the distribution

$N(\mathbf{0}, \sigma^2 I - \frac{\sigma^4}{(n\sigma^2+\tau^2)} ee')$, which is monotically decreasing in τ^2 (compare with the example in Section 3.2.4).

3.3.4 *Negative association*

Joag-Dev and Proschan [106] use the simple idea that a negative dependence means that splitting the set of random variables into two parts leads to a negative covariance between the two sets.

Definition : Random variables $(X_1, X_2, ..., X_k)$ are negatively associated (NA), if for every pair A_1, A_2 of disjoint subsets of $\{1, 2, ..., k\}$ and whenever f_1 and f_2 real-valued increasing functions

$$Cov(f_1(X_i, i \in A_1), f_2(X_j, j \in A_2)) \leq 0 .$$

Note that the condition can be replaced by $f_1(X_i, i \in A_1)$ and $f_2(X_j, j \in A_2)$ are NQD.

Remark that there is no symmetry between this definition and the definition of positive association (Section 3-2-3). Namely, even for two variables, the property

$$Cov(f(X,Y), g(X,Y)) \leq 0 \qquad \forall f, \forall g \text{ increasing functions}$$

is impossible if $f \equiv g$. The NA class of distributions enjoys a number of appealing properties, in particular it is closed under formation of increasing functions of disjoint sets of random variables. These properties imply an important theorem on the distribution of the joint conditional distribution of k independent random variables given their sum. Using this theorem, it is easy to prove that many distributions are NA, in particular multinomial, multivariate hypergeometric and Dirichlet distributions. We shall enumerate the properties and state the theorem.

(1) P1 : For a pair of random variables the NQD property is equivalent to the NA property.

(2) P2 : Let $A_1, A_2, ..., A_n$ be disjoint subsets of $\{1, 2, ..., k\}$, and $f_1, f_2, ...$, be increasing positive functions. Then if $X_1, X_2, ..., X_k$ are NA :

$$E\Pi_{i=1}^n f_i(X_j, j \in A_i) \leq \Pi_{i=1}^n E f_i(X_j, j \in A_i).$$

(3) P3 :

P2 property implies that, if A_1 and A_2 are any disjoint subsets of $\{1, 2, ..., k\}$, and $x_1, ..., x_k$ are real values, then

$$P(X_i \le x_i,\ i = 1, 2, ..., k) \le P(X_i \le x_i, i \in A_1)P(X_j \le x_j, j \in A_2)$$

and

$$P(X_i > x_i,\ i = 1, 2, ..., k) \le P(X_i > x_i, i \in A_1)P(X_j > x_j, j \in A_2)\ .$$

In particular $X_1, ..., X_k$ are NUOD and NLOD.

(4) P4 : A subset of two or more NA random variables is NA.

(5) P5 : A set of independent random variables is NA.

(6) P6 : Increasing functions defined on disjoint subsets of a set of NA random variables are NA.

(7) P7 : The union of independent sets of NA random variables are NA.

Remark: Neither NUOD nor NLOD imply NA. Joag-Dev and Proschan [106] show it with the following counterexample :

Table 3.1 *A vector (X_1, X_2, X_3, X_4) of binary random variables, which is NUOD and NLOD but not NA*

		(X_1, X_2)				
		(0,0)	(0,1)	(1,0)	(1,1)	bivariate marginals
(X_3, X_4)	(0,0)	.0577	.0623	.0623	.0577	.24
	(0,1)	.0623	.0677	.0677	.0623	.26
	(1,0)	.0623	.0677	.0677	.0623	.26
	(1,1)	.0577	.0623	.0623	.0577	.24
	bivariate marginals	.24	.26	.26	.24	

The marginal distributions of X_i, $i = 1, ..., 4$ are binary. Here the low probabilities in the "tails" of the distribution $(0, 0, 0, 0)$ and $(1, 1, 1, 1)$ result that the NLOD and NUOD conditions are satisfied. However the property P3 is not valid due to the fact that the probabilities are high in the center of the distribution, namely:

$$P(X_i = 1,\ i = 1, ..., 4) > P(X_1 = 1,\ X_2 = 1)P(X_3 = 1,\ X_4 = 1)\ .$$

Theorem [106] : let $X_1, X_2, ..., X_k$ be k independent random variables with PF_2 densities. Then the joint conditional distribution of $X_1, X_2, ..., X_k$ given their sum ΣX_i is NA.

Application : Multinomial, multivariate hypergeometric, Dirichlet distributions are all NA.

Remark : The NA property does not imply RR_2 in pairs property as the following rather artificial example shows. Let $X = (X_1, X_2, X_3)$ be a random vector having a trivariate multinomial frequency function f with strictly positive probabilities p_1, p_2, p_3 and satisfying $X_1 + X_2 + X_3 = 3$. Consider the vector $Y = (Y_1, Y_2)$, where $Y_1 = X_1 X_2$ and $Y_2 = X_3$, and let g be the frequency of (Y_1, Y_2). Since f is multinomial, it is RR_2 in pairs, but g does not inherit this property. In particular, one can verify that $g(0,0)g(1,1) > g(0,1)g(1,0)$. However f being trinomial, it is NA, and therefore by property P6, g is NA.

Theorem [104] : Let $X_1, X_2, ..., X_k$ be k independent random variables with continuous distribution. Then the joint conditional distribution of $(X_1, X_2, ..., X_k)$ given the joint order-statistic $(X_{(k_1)} = s_1, X_{(k_2)} = s_2, ..., X_{(k_r)} = s_r)$ is NA, for any $1 < k_1 < k_2 < ... < k_r < k$.

Theorem [104] : Let $X_1, X_2, ..., X_k$ be k independent random variables with continuous distribution. Then the joint conditional distribution of $X_1, X_2, ..., X_k$ given the joint order- statistic $(X_{(k_1)} < s_1, X_{(k_2)} < s_2, ..., X_{(k_r)} < s_r)$ is NA, for any $1 < k_1 < k_2 < ... < k_r < k$.

Theorem : Negatively correlated normal random variables are NA.

Let $\mathbf{X} = (X_1, ..., X_n)$ be a multivariate normal random vector, with $corr(X_i, X_j) = \rho_{ij}$, $1 \leq i < j \leq n$. Then $\mathbf{X}$ is positively associated if ρ_{ij} is positive and NA if ρ_{ij} is negative. See [107] and [106] for a proof.

3.3.5 *Negatively superadditive dependence*

Hu [103] has used the supermodular functions to define negative dependence by reversing the properties shown in Section 3.2.2 on positive dependence.

Definition : A random vector $\mathbf{X} = (X_1, X_2..X_n)$ is said to be negatively superadditive dependent (NSD) if

$$E\Phi(\mathbf{X}) \leq E\Phi(\mathbf{Y})$$

where $\mathbf{Y} = (Y_1, Y_2, ..., Y_n)$ is a vector of independent variables having the same univariate marginal distributions as $\mathbf{X}$, and for all $\Phi : R^m \mapsto R$ supermodular.

The NSD dependence has the following properties :

(1) P1. For a pair of random variables NSD is equivalent to NQD. This is obtained using the same arguments as in corollary, Section 3.2.2 concerning PQD.
(2) P2. NSD implies NUOD and NLOD. The reasoning is the same as in Section 3.2.2.
(3) P3. If $\mathbf{X} = (X_1, X_2, ..., X_n)$ is NSD and g_i , $i = 1, ..., n$ are increasing functions, then $(g_1(X_1), ..., g_n(X_n))$ is NSD.
(4) P4. The property of NSD is kept by permutation.
(5) P5. If $\mathbf{X} = (X_1, X_2, ..X_n)$ is NSD, then all marginals of this distribution are NSD.
(6) P6. If $\mathbf{X} = (X_1, X_2, ..X_n)$ and $\mathbf{Y} = (Y_1, Y_2, ..Y_n)$ are NSD and independent of each other, then $(\mathbf{X}, \mathbf{Y})$ is NSD.
(7) P7. If $\mathbf{X} = (X_1, X_2, ..., X_n)$ and $\mathbf{Y} = (Y_1, Y_2, ..Y_n)$ are NSD and independent of each other, then $\mathbf{X}+\mathbf{Y} = (X_1+Y_1, X_2+Y_2, ..., X_n+ Y_n)$ is NSD.
(8) P8. If $\mathbf{X} = (X_1, X_2, ..., X_n)$ is NSD , and X_i and X_j are uncorrelated for all $i \neq j$, then $X_1, X_2, ...X_n$ are mutually independent.

Property P8 sheds an additional light on the relation between $\rho = 0$ and independence.

NSD does not imply NA, and the converse is, to the best of our knowledge, an open problem. Hu [103] using the same example as Joag-Dev and Proschan (Table 3.1, Section 3.3.4) shows that the distribution constructed by these authors using binary variables, which is not NA, is NSD.

For NSD property, we have a theorem analogous to the theorem concerning Pólya functions and conditioning by the sum of the random variables to the case of NA vectors:

Theorem: Let $X_1, X_2, ..., X_k$ be k independent random variables each having PF_2 density. Then the joint conditional distribution of $X_1, X_2, ..., X_k$ given ΣX_i is NSD.

All the examples of distributions presented in the preceding sections which are obtained from conditioning by a sum of Pólya functions are NSD: multinomial, multivariate hypergeometric, Dirichlet, Dirichlet compound multinomial.

Another example, proposed by Block *et al.* [23] of NSD distribution is a special case of multivariate Farlie-Gumbel-Morgenstern (FGM) distribution (see Chapter 5) given by :

$$F(\mathbf{x}) = \Pi_{i=1}^{n} (1 + \Sigma_{i<j}\theta_{ij}(1 - F_i(x_i))(1 - F_j(x_j)))$$

where $\theta_{ij} < 0$ and such that the density is positive [23].

This dependence seems to have applicability in studies related to coherent systems [24].

3.4 Setwise Dependence

Many concepts of positive dependence have been extended to setwise positive dependence by Chhetry *et al.* [43].

- A p-dimensional vector $\mathbf{X} = (X_1, ..., X_p)$ can be partitioned into k ($k \geq 2$) subvectors $\mathbf{X}_1$, $\mathbf{X}_2$,...,$\mathbf{X}_t$,...,$\mathbf{X}_k$ of dimensions p_1,p_2,...,p_k ($\Sigma_{i=1}^{k} p_i = p$) respectively, namely we define a partition $C = \{C_1, ..., C_k\}$ of the index set $1, 2, ..., p$. We call p_t the dimension of the vector $\mathbf{X}_t$. All the notions of setwise dependence are presented relative to a *fixed partition* C.

3.4.1 *Setwise upper orthant and setwise upper set positive dependences*

Definition 1: The vector $\mathbf{X}$ with a partition C into k parts is setwise positively upper orthant dependent (SPUOD) if for all $\mathbf{X} = (\mathbf{X}_1, \mathbf{X}_2, ..., \mathbf{X}_k)$,

$$P\left(\bigcap_{t=1}^{k}\{\mathbf{X}_t > \mathbf{x}_t\}\right) \geq \Pi_{t=1}^{k} P(\mathbf{X}_t > \mathbf{x}_t).$$

The definition of positively lower orthant dependence SPLOD is obtained by reversing the inequalities in the expression of probabilities.

Definition 2 : Replacing the orthant $]\mathbf{x}_t, \infty[$ by an upper set U_t we obtain a definition of setwise positively upper set dependence SPUSD, and similarly we can define SPLSD from SPLOD.

$$P\left(\bigcap_{t=1}^{k}\{\mathbf{X}_t \in U_t\}\right) \geq \Pi_{t=1}^{k} P(\mathbf{X}_t \in U_t).$$

If, in the preceding definitions we replace the inequality by an equality, we then have the definition of setwise independence (SI), as a particular case of setwise dependence.

One observes that SPUSD (SPLSD) implies SPUOD (SPLOD), but the converse is false. If Pr designates one of these four concepts, we then have

(1) P1
Let $\mathbf{X}_t^*$ be a subvector of $\mathbf{X}_t$; if $\mathbf{X}_1, \mathbf{X}_2, ..., \mathbf{X}_k$ satisfy P, then $\mathbf{X}_1^*$, $\mathbf{X}_2^*, ..., \mathbf{X}_k^*$ satisfy Pr;
(2) P2
If $\mathbf{X}_1, \mathbf{X}_2, ..., \mathbf{X}_k$ and $\mathbf{Y}_1, \mathbf{Y}_2, ..., \mathbf{Y}_r$ satisfy Pr, so do $\mathbf{X}_1, \mathbf{X}_2, ..., \mathbf{X}_k, \mathbf{Y}_1, \mathbf{Y}_2, ..., \mathbf{Y}_r$;
(3) P3
If $\mathbf{X}_1, \mathbf{X}_2, ..., \mathbf{X}_t, ..., \mathbf{X}_k$ satisfy SPUSD (or SPLSD), then for every set of increasing functions in each component $h_t : R^{p_t} \mapsto R^{q_t}$, $t = 1, ..., k$, $h_1(\mathbf{X}_1), h_2(\mathbf{X}_2), h_t(\mathbf{X}_t), ..., h_k(\mathbf{X}_k)$ satisfies SPUSD (or S-PLSD).
(4) P4
Let $\mathbf{X}^n = (X_1^n, ..., X_p^n)$ be a sequence of random multivariate vector. If for a fixed partition, the sequence $\mathbf{X}^n$ satisfies Pr for all n and if $\mathbf{X}^n$ converges in distribution to $\mathbf{X}$, then $\mathbf{X}$ satisfies Pr w.r.t the same partition.

When $k = 2$ the concepts of SPUSD and SPLSD are equivalent, but this equivalence is not valid for SPUOD and SPLOD which are not complementary sets.

Theorem : Let $(\mathbf{X}_1', \mathbf{X}_2', ..., \mathbf{X}_k')'$ be a normal random vector with mean zero and covariance matrix Σ . Denote $\Sigma_{ij} = cov(\mathbf{X}_i, \mathbf{X}_j)$, the covariance matrix of $(\mathbf{X}_i, \mathbf{X}_j)$ then, $\mathbf{X}_1$, $\mathbf{X}_2$,...,$\mathbf{X}_k$ are SPUSD if and only if Σ_{ij} is positive or identically null (i.e. all its elements are positive or identically zero) for all $i \neq j$.

For multivariate normal vectors, all the properties SPUSD, SPLSD, SPUOD, SPLOD are equivalent to the property $cov(\mathbf{X}_i, \mathbf{X}_j) \geq 0$.

Theorem (Chhetry *et al.* [44]) : If the two random vectors $\mathbf{X}_1$ and $\mathbf{X}_2$ are SPUSD and satisfy $cov(\mathbf{X}_1, \mathbf{X}_2) = 0$, they are SI. However $\mathbf{X}_1$ and $\mathbf{X}_2$ being SPUOD and $cov(\mathbf{X}_1, \mathbf{X}_2) = 0$ do not guarantee that $\mathbf{X}_1$ and $\mathbf{X}_2$ are SI.

3.4.2 *Setwise association*

The notion of association can be extended to a setwise association (SA).

Definition: Random vectors $\mathbf{X}_1$, $\mathbf{X}_2$,..., $\mathbf{X}_k$ are setwise associated if:

$$cov\,(f_1(a_1(\mathbf{X}_1),...,a_k(\mathbf{X}_k)),f_2(a_1(\mathbf{X}_1),...,a_k(\mathbf{X}_k))) \geq 0$$

for all increasing functions $a_t \ : R^{p_t} \mapsto R,\ t = 1,...,k$ and for every pair of increasing functions $f_1,\ f_2\ :\ R^k \mapsto R$.

$\mathbf{X}$ is associated if it is setwise associated with respect to all possible partitions.

The properties P1 to P4 in the preceding section remain valid if we replace Pr by SA.

The condition SA is stronger than that of SPUSD and that of SPLSD (and consequently stronger than SPUOD and SPLOD).

Theorem: Let $\mathbf{X}_1$,...$\mathbf{X}_k$ be conditionally independent given a vector $\mathbf{W}$; if each $\mathbf{X}_t$ is stochastically increasing in $\mathbf{W}$ and if $\mathbf{W}$ is associated, then $\mathbf{X}_1$,...,$\mathbf{X}_k$, $\mathbf{W}$ are SA and in particular $\mathbf{X}_1$,...,$\mathbf{X}_k$ are SA.

Corollary: Let $\mathbf{U} = \mathbf{U}_1{}'$,...,$\mathbf{U}_k{}'$ be SI , and a random vector $\mathbf{Z} = (\mathbf{Z}_1', \mathbf{Z}_2', ..\mathbf{Z}_k')$ be associated, and $\mathbf{U}$ and $\mathbf{Z}$ be SI, then $\mathbf{U}_1 + \mathbf{Z}_1$, $\mathbf{U}_2 + \mathbf{Z}_2$, ...$\mathbf{U}_k + \mathbf{Z}_k$ are SA.

An application to a multivariate normal vector : Let $\mathbf{X} = (\mathbf{X}_1{}', ...\mathbf{X}_k{}')'$ be distributed as $N(0,\Sigma)$ where $\Sigma = Diag(\Sigma_1,...,\Sigma_k)+\Lambda$ with Σ_i, $i = 1...k$ and Λ are non-negative definite matrices and Λ having all its elements positive or zero, then $\mathbf{X}$ is SA.

The proof uses the fact that any multivariate normal random vector $\mathbf{U} = (\mathbf{U}_1{}', ...\mathbf{U}_k{}')'$ with mean zero and covariance matrix $Diag(\Sigma_1,...,\Sigma_k)$ is SI, and that any $\mathbf{Z} = (\mathbf{Z}_1{}', ..., \mathbf{Z}_k{}')'$ independent of $\mathbf{U}$ with the distribution $N(0,\Lambda)$ is associated (in the sense of Section 3.2.3), hence $\mathbf{U}_1+\mathbf{Z}_1$, $\mathbf{U}_2+\mathbf{Z}_2$, ...$\mathbf{U}_k + \mathbf{Z}_k$ are SA.

3.4.3 *Setwise dependence by mixture*

The concept of positive dependence by mixture (Section 3.2.8) can be extended to setwise dependence by mixture. A random vector $\mathbf{X} = (\mathbf{X}_1{}', ..., \mathbf{X}_t{}', ..., \mathbf{X}_k{}')'$ (where the $\mathbf{X}_t$ have the same dimension p) is setwise dependent by mixture (SDM) if there exists a random vector $\mathbf{W}$ with the distribution function G such that conditionally to W the vectors $\mathbf{X}_t$ are SI and identically distributed with distribution function F^w. The distribution function of $\mathbf{X}$ is then:

$$F(\mathbf{x}) = \int \Pi_{i=1}^k F^{\mathbf{w}} dG(\mathbf{w}).$$

Some basic properties are:

(1) If $\mathbf{X} = (\mathbf{X}_1', ..., \mathbf{X}_k')'$ is SDM, then for any vector $\mathbf{a} = (\mathbf{a}_1', ..., \mathbf{a}_k')'$ in R^p

$$P(\mathbf{X}_i > \mathbf{a}) \geq \Pi_{i=1}^k P(\mathbf{X}_i > \mathbf{a}_i)$$

and

$$P(\mathbf{X}_i \leq \mathbf{a}) \geq \Pi_{i=1}^k P(\mathbf{X}_i \leq \mathbf{a}_i).$$

(2) If $\mathbf{X} = (\mathbf{X}_1', ..., \mathbf{X}_k')'$ is SDM, then $\Sigma_{ij} = cov(\mathbf{X}_i, \mathbf{X}_j)$ is non-negative definite for all i and j.
To prove this one can use the so called second Chebyshev's identity:

$$Cov(\mathbf{X}_i, \mathbf{X}_j) = E(Cov(\mathbf{X}_i, \mathbf{X}_j)/\mathbf{W}) + cov(E(\mathbf{X}_i/\mathbf{W}), E(\mathbf{X}_j/\mathbf{W})).$$

(3) Corollary: Let $\mathbf{X} = (\mathbf{X}_1', ..., \mathbf{X}_t'..\mathbf{X}_k')'$, a standardized multivariate normal random vector, be distributed as $N(0, \Sigma)$, with $\Sigma_{ii} = \Sigma_1$ for $i = 1, ..., k$ and $\Sigma_{ij} = \Sigma_2$ for $i \neq j$, then $\mathbf{X}$ is SDM if and only if Σ_2 is non-negative definite.

3.4.4 *Extension to the setwise negative dependence*

For negative dependence it is straightforward to define the concepts of S-NUOD, SNLOD, SNUSD, SNLSD by reversing the definitions of SPUOD, SPLOD, SPUSD, SPLSD. When $k = 2$, if $\mathbf{X}$ is SNUSD or SNLSD for all the partitions, then $\mathbf{X}$ is NA. In the same manner as in the case of positive dependence upper set dependence is stronger than upper orthant dependence and correspondingly the lower set stronger than lower orthant one (see Chhetry *et al.* [43] for details). Other concepts of negative dependence are more difficult to extend to setwise dependence.

3.5 Other Approaches

In this chapter, our approach has been to study the connection between dependence/independence with uncorrelatedness and show that many known distributions and order statistics have NA dependence and that conditioning creates NA. We have not really considered the approach of Lehmann, that is inferences problems and test issues. Many authors have tackled these problems, for example, Roussas [185] [184] have studied non-parametric estimation of survival function and quantiles, density or hazard rate estimation with a kernel under positive or negative association. Cohen *et al.* [49] have used the notion of cone order association, useful for establishing unbiasedness of test parameters. Experimental studies describing various methods for assessing dependence are presented in Clemen *et al.*

3.6 Positive Dependence Orderings

When a concept of dependence is defined for two variables, we can compare a bivariate distribution F with the distribution F_1F_2 obtained under the assumption of independence. More generally, we wish to compare two bivariate distributions F and G *with the same marginals*, to find out whether one distribution is more positively dependent than the other. Many authors, Yanagimoto and Okamoto [226], Capéraà and Genest [37], Kimeldorf and Sampson [130], Rinott and Pollack [182], Schriever [193], have defined orderings based on PQD, PRD, association, positive dependence function DTP(0,1) and LRD. Other orderings have been defined by Shaked and Tong [203] and Scarsini and Venetoulias [191]. Here, we present orderings defined for bivariate distributions, but it is possible to extend these orderings to compare multivariate distributions. All the distributions under comparison have the same marginals, and this makes relevant to investigate the dependence structure of the random vectors.

On the other hand, it is also of interest to compare random vectors with different marginals, but the same dependence structure. In the last section we provide a few elements about *integral stochastic orderings* . Chang [41], Shaked and Shantikumar [205] *et al.*, Scarsini [192], Bäuerle [18], Müller [159] [158] among others have studied these orderings thoroughly.

3.6.1 *Ordering based on PQD*

The simplest way to generalize the PQD ordering of two distributions F and G is to stipulate that:

$$F \ll G \text{ if } F(x,y) \leq G(x,y) \; \forall (x,y).$$

Namely, if the distribution $F = G_1 \cdot G_2$, corresponds to the hypothetical independence of the distribution G, we retrieve the PQD property for the distribution G. The relation $\ll$ is named concordance ordering. This is the weakest ordering that can be defined.

3.6.2 *Conditions on ordering*

In order to propose orderings stronger than concordance ordering, Kimeldorf and Sampson [130] have stipulated nine conditions that a positive dependence ordering $\ll$ has to satisfy for a family of distributions $\mathcal{F}(F_1, F_2)$ with the same marginals F_1 and F_2. The first condition is the concordance ordering, next we have to verify the conditions of reflexivity, antisymmetry and transitivity. The other conditions assert that the ordering is compatible with the upper and lower bounds for the family, and with the convergence in distribution and is invariant with respect to monotone transformations and exchangeability.

To summarize, $\ll$ is an ordering on the family $\mathcal{F}(F_1, F_2)$ if it satisfies:

(1) concordance :
$F \ll G$ implies that $F(x,y) \leq G(x,y)$ for all x and y ;

(2) reflexivity
$F \ll F$ for all F ;

(3) antisymmetry :
$F \ll G$ and $G \ll F$ implies $F = G$;

(4) transitivity :
$F \ll G$ and $G \ll H$ implies $F \ll H$;

(5) The Fréchet bounds belong to the family, and $F^- \ll F \ll F^+$ where as before $F^+ = min(F_1, F_2)$ and $F^- = max(F_1 + F_2 - 1, 0)$

(6) Closure with respect to convergence in distribution :
$F_n \xrightarrow{L} F$, $G_n \xrightarrow{L} G$ then $F_n \ll G_n$ for all n implies $F \ll G$

(7) Invariance with respect to increasing transformation :
If Φ is an increasing function, then $(X,Y) \ll (U,V)$ implies $(\Phi(X), Y) \ll (\Phi(U), V)$;

(8) $(X,Y) \ll (U,V)$ implies $(-U,V) \ll (-X,Y)$;

(9) exchangeability :
$(X,Y) \ll (U,V)$ implies $(Y,X) \ll (V,U)$.

3.6.3 *Ordering defined by PRD*

Yanagimoto and Okamoto [228] have proposed an ordering defined by PRD. To compare the two bivariate distributions F and G, we use the distributions $F_{x',x} = F_{Y/x'}F_{Y/x}^{-1}$ and $G_{x',x} = G_{Y/x'}G_{Y/x'}^{-1}$, defined in Section 3.2.5. When $F_{Y/x}$ and $G_{Y/x}$ are continuous and strictly increasing, we define the ordering in the following manner :

$$F \ll_{PRD} G\ if\ \forall x < x',\ \forall\, 0 \leq u \leq 1\ F_{x',x}(u) \geq G_{x',x}(u). \tag{3.13}$$

The equation above implies that the distribution G is more concentrated than the distribution F. We can verify that if F corresponds to the independence situation for G, then $F_{x,x'}(u) = u$, and we retrieve the proposition given in Section 3.2.5 for PRD.

Note that this ordering is not symmetric in x and y and consequently we cannot verify the axiom of exchangeability.

3.6.4 *Association ordering*

The notion of association of variables prompted Schriever [193] to introduce the following order $\ll_{ass}$:

Definition: Let (X, Y) and (W, Z) be two random bivariate vectors having the distributions F and G respectively. Then :

$$F \ll_{ass} G \text{ if}$$
$$\exists\, K_1,\ \exists\, K_2 \text{ increasing functions in both arguments, such that}$$
$$(X, Y) \stackrel{L}{=} (K_1(W, Z), K_2(W, Z))$$

where K_1 and K_2 satisfy :

$$K_1(x_1, y_1) < K_1(x_2, y_2)\ , K_2(x_1, y_1) > K_2(x_2, y_2) \rightarrow\ x_1 < x_2\ , y_1 > y_2\ . \tag{3.14}$$

This new condition is required to assure that the $\ll_{ass}$ ordering will be applicable when the variables (X, Y) are not exchangeable.

Properties of this ordering are:

- Association-ordering is invariant under scale and location transformations.

- PRD-ordering implies association-ordering. Namely, if $F \ll_{PRD} G$, then it suffices to take $K_1(x, y) = x$ to arrive at $F \ll_{ass} G$.

Example: Let X and Y be two random variables and define $X^\alpha = (1 - \alpha X) + \alpha Y$, $Y^\alpha = \alpha X + (1 - \alpha)Y$ with $\alpha \in [0, 1/2]$, then :

$$\forall 0 \leq \alpha_1 < \alpha_2 \leq 1/2 \ , \ (X^{\alpha_1}, Y^{\alpha_1}) \ll_{ass} (X^{\alpha_2}, Y^{\alpha_2}).$$

This example has been used by several authors, see e.g. Schriever [193] and Yanagimoto [226].

3.6.5 *PDD-ordering*

Let (X, Y), (W, Z), be two random vectors with distribution function F and G. Utilizing the concept of positive function dependence (Section 3.2.4), Rinott and Pollack [182] introduce the ordering $\ll_{PDD}$ as follows:

$$F \ll_{PDD} G \quad if\ Cov(h(X), h(Y)) \leq Cov(h(W), h(Z))$$

for every real function h. Since the two distributions have the same marginals, this is equivalent to

$$E(h(X)h(Y)) \leq E(h(W)h(Z))$$

for every real function h. It is then possible to extend this ordering to the case of two random vectors $\mathbf{X} = (X_1, X_2, ..., X_n)$ and $\mathbf{Y} = (Y_1, Y_2, ..., Y_n)$ $(n > 2)$:

$$X \ll_{PDD} Y \text{ if } E(\Pi_{i=1}^n h(X_i)) \leq E(\Pi_{i=1}^n h(Y_i))$$

Rinott and Pollack established the following characteristic property:

$$F \ll_{PDD} G \Leftrightarrow F(x, y) - G(x, y) \text{ is a positive definite kernel.}$$

The PDD-ordering is preserved under mixture and under limits in distribution.

The preservation under mixture means that: if (X, Y, θ) and (W, Z, θ) are two random vectors depending on a parameter $\theta \in \Theta$, then

$$\forall \theta \in \Theta,\ [(X, Y)/\Theta = \theta] \ll_{PDD} [(W, Z)/\Theta = \theta] \Rightarrow (X, Y) \ll_{PDD} (W, Z)\,.$$

3.6.6 *Orderings defined from DTP(0,1) and LRD*

In the same manner as in Section 3.6.3, we can generalize the inequality (3.10), which compares the ratio $\frac{S_{x'x}(u)}{S_{x'x}(v)}$ with the ratio obtained (under independence) from the uniform survival function $\frac{1-u}{1-v}$, for comparing two distributions as follows :

$$F \ll_{DTP(0,1)} G \quad if\, \forall x < x' \, \forall\, 0 \leq u < v < 1 \; \frac{S^G_{x'x}(u)}{S^G_{x'x}(v)} \leq \frac{S^F_{x'x}(u)}{S^F_{x'x}(v)}$$

(where the exponent in the survival function indicates the associated distribution function).

To define the DTP(0,1)-ordering, Capéraà and Genest require additionally that :

$$\frac{G_{x'x}(u)}{G_{x'x}(v)} \leq \frac{F_{x'x}(u)}{F_{x'x}(v)}.$$

Equivalently, one can assert that the ratio $\frac{G_{x'x}}{F_{x'x}}(u)$ with $x < x'$ is an increasing function of u.

The LRD-dependence is expressed as :

$$F \ll_{LRD} G \quad \text{if } \forall x < x', \quad \forall\, 0 \leq u < t < v < 1$$
$$\frac{G_{x'x}(t) - G_{x'x}(u)}{G_{x'x}(v) - G_{x'x}(u)} \leq \frac{F_{x'x}(t) - F_{x'x}(u)}{F_{x'x}(v) - F_{x'x}(u)}. \tag{3.15}$$

If we suppose that the c.d.f $F_{x'x}$ and $G_{x'x}$ have densities, we can rewrite this last inequality as

$$\frac{dG_{x'x}(u)}{dF_{x'x}(u)} \tag{3.16}$$

is an increasing function of u, when $x < x'$.

The preceding generalization is due to Capéraà and Genest in [37]. Kimeldorf and Sampson [129] have defined another TP2-ordering ($\ll_{TP2}$) as follows :

Let $I \times J$ be a rectangle, and $F(I, J)$, $G(I, J)$ be the associated probabilities . We write $I_1 < I_2$, if $\forall x \in I_1$ and $\forall y \in I_2$, $x < y$.

$$F \ll_{TP2} G \quad \text{if} \quad \forall I_1 < I_2, \forall J_1 < J_2 ,$$

$$F(I_1, J_1)F(I_2, J_2)G(I_1, J_2)G(I_2, J_1) \leq F(I_1, J_2)F(I_2, J_1)G(I_1, J_1)G(I_2, J_2). \tag{3.17}$$

This ordering is different from the LRD-ordering suggested by Capéraà and Genest [37] and does not imply the DTP(0,1)- and the PRD-ordering as the LRD-ordering does. (Capéraà and Genest provide a counterexample to show this non-implication.)

Proposition: Let F and G be two distribution functions, with the same marginals and such that the conditional distribution $F_{Y/x}$ and $G_{Y/x}$ have supports independent of x, then :

$$F \ll_{LRD} \Rightarrow F \ll_{DTP(0,1)} G \Rightarrow F \ll_{PRD} G.$$

Examples:

(1) Yanagimoto and Okamoto's example [227]
Let $X = U$, $Y_\alpha = (1+\alpha U)V$ with $\alpha > -1$ and U and V be independent random variables with an absolutely continuous distribution. Yanagimoto and Okamoto have shown that this family is ordered by PRD-ordering, and Capéraà and Genest [37] have proved that it is ordered by LRD-ordering.

(2) Frank's Copula (see chapter 4, Section 4.6.2) The copula is defined by :

$$F_\alpha(x,y) = log_\alpha(1 + \frac{(\alpha^x - 1)(\alpha^y - 1)}{\alpha - 1}) \qquad 0 < \alpha < 1$$

and for $\alpha = 1$, $F_1(x,y) = xy$ for $0 \leq x, y \leq 1$
This family possesses the LRD-ordering , namely one can verify that for $x < x'$, $\alpha' < \alpha$ the ratio

$$\frac{f_{\alpha,x',x}(u)}{f_{\alpha',x',x}(u)}$$

is increasing with u, which is the characteristic property of LRD-ordering defined above (relation 3.16).

(3) Clayton's Copula (see Chapter 4, Section 4.6.2) Here the c.d.f is

$$F_\alpha(x,y) = (x^{-\alpha} + y^{-\alpha} - 1)^{-1/\alpha},\ \alpha > 0,\ 0 \leq x, y \leq 1\,.$$

Denote $F_{\alpha,x',x}(u) = F_{\alpha,x'} F_{\alpha,x}^{-1}(u)$ with $x < x'$. This family is not ordered by the DTP(0,1)-ordering, namely the ratio $\frac{F_{\alpha',x',x}(u)}{F_{\alpha,x',x}(u)}$, with $\alpha < \alpha'$, is not an increasing function of u.

3.6.7 *Integral stochastic orderings*

In this section we do not impose that the two vectors to be compared have the same marginals. Consider a random vector $\mathbf{X}$ in R^n for $n \geq 2$. The orderings to be defined are called *integral stochastic orderings*, because they are defined for a class of real-valued functions $\mathcal{F}$, such that $\mathbf{X} \ll \mathbf{Y}$ holds if and only if $E(f(\mathbf{X})) \leq E(f(\mathbf{Y}))$ for all f belonging to $\mathcal{F}$ (for which the expectations exist).

For example $\mathcal{F}$ can be the set of all *component-wise increasing* functions, or the set of all *convex functions* , or all *supermodular* , or all *componentwise convex*, or all *directionally convex* functions. (A function is directionally convex if it is supermodular and componentwise convex.)

We have already seen in Section 3.2.2 that PQD-dependence is closely related to certain properties on supermodular and componentwise increasing functions.

We shall give here a few properties of supermodular ordering, which show that this ordering is suitable for the comparison of distributions with the same marginals. Later we discuss directionally convex ordering as an ordering for vectors with different variability in the marginals.

3.6.7.1 *Supermodular ordering*

(1) $\mathbf{X} \ll_{sm} \mathbf{Y}$ implies $P(\mathbf{X} > \mathbf{x}) \leq P(\mathbf{Y} > \mathbf{y})$ and $P(\mathbf{X} \leq \mathbf{x}) \leq P(\mathbf{Y} \leq \mathbf{y})$. Thus supermodular-ordering implies upper-orthant-ordering and lower-orthant-ordering (where upper and lower orthant orderings are natural extensions of PQD-ordering).
(2) $\mathbf{X} \ll_{sm} \mathbf{Y}$ implies $X_i \stackrel{L}{=} Y_i$ for all $i = 1, ..., n$. Therefore the supermodular ordering presupposes that the two vectors have the same univariate marginals and therefore compares only their dependence structure (i.e. their copulas, see Chapter 4).
(3) $\mathbf{X} \ll_{sm} \mathbf{Y}$ implies $cov(f(X_i), g(X_j)) \leq cov(f(Y_i, g(Y_j))$, for all f and g $R \mapsto R$ both increasing or decreasing and for all i and j in $1, 2, ..., n$.
(4) The Lorentz inequality. Let $(X_1, X_2, ..., X_n)$ be identically distributed random variables. Then
$F \ll_{sm} F^+$, where F^+ is the upper Fréchet bound.
(5) Connection with the CIS-dependence:
If $\mathbf{X}$ is CIS and $\mathbf{X}^*$ is an independent vector with the same marginals

as $\mathbf{X}$, then $\mathbf{X}^* \ll_{sm} \mathbf{X}$.

3.6.7.2 *Directionally convex ordering*

Sometimes it is of interest to compare some positive convex combinations of the components of the two random vectors $\mathbf{X}$ and $\mathbf{Y}$. If the two random vectors have a common copula, then if their components are ordered by the increasing-ordering ($Eg(X_i) \ll Eg(Y_i)$) for all i and for all g increasing functions), this is also true for the vectors (see Scarsini [190]). But if the components are ordered just by the convex order (g is convex), then this is not true for the vectors [190]. The idea of Müller and Scarsini [159] is then to use the *directionally convex* functions which combine the properties of convexity of the components and the supermodularity property of the function to construct an ordering. They show the following two properties:

- If $\mathbf{X}$ and $\mathbf{Y}$ have a common copula which is CI (conditionally increasing) and if $Eg(X_i) \leq Eg(Y_i)$ for all convex-functions $g : R \mapsto R$, and all $i = 1, 2, ..., n$ then $Ef(\mathbf{X}) \leq Ef(\mathbf{Y})$ for all directionally convex functions $f : R^n \mapsto R$.
- If $Ef_i(X_i) \leq Ef_i(Y_i)$ for all convex functions f_i implies $Ef(\mathbf{X}) \leq Ef(\mathbf{Y})$ for all comonotone random vectors $\mathbf{X}$ and $\mathbf{Y}$, then f is directionally convex. (A vector $\mathbf{X}$ is comonotone if its distribution is the upper Fréchet bound.)

3.6.8 *Generating a family of partial orderings*

Yanagimoto [228] followed by Metry and Sampson [155] provided an approach to unify the TP_2-ordering and the concordance ordering as well as to generate other orderings. The idea is to adjust the endpoints of the intervals of the rectangles $I_1 \times J_1$ and $I_2 \times J_2$ utilized in the definition of the TP_2-ordering. An endpoint of any interval can be $\pm\infty$, and the two intervals I_1 and I_2 (or J_1 and J_2) may or may not possess a non-empty intersection.

In this manner Metry and Sampson generate 64 orderings, of which concordance ordering and TP_2-ordering are particular cases. More precisely, with the following notations :

- L is the set of left intervals : $]-\infty, a[$ or $]-\infty, a]$
- R is the set of right intervals :$[a, +\infty[$ or $]a, +\infty[$
- S is the set of any arbitrary intervals ($L \subset S$ and $R \subset S$)

- If two disjoint intervals share a common endpoint $I_1 \leq I_2$, the corresponding notation will be A (A stands for "abut"), and NA if they (not necessarily) share a common endpoint (the two disjoint intervals are not required to "abut", but may do so) . For example the concordance ordering is represented by (L,R; L,R; A, A), and the TP_2-ordering by (S, S; S, S; NA,NA). (S,S;S,S;A,A) corresponds to an ordering defined as P(3,3)-ordering by Yanagimoto [228]. Thus, using this notation, we have :

$$\begin{array}{ccccc} & & (.,.;.,.;NA,A) & & \\ & \nearrow & & \searrow & \\ (.,.;;,.;NA,NA) & & & & (.,;.,;A,A) \\ & \searrow & & \nearrow & \\ & & (.,;.,;A,NA) & & \end{array}$$

where any combination of L, R and S may be substituted for the first pair ".,.". (The two arrows $\nearrow$ and $\searrow$ mean "implies".)

Moreover,

$$\begin{array}{ccccc} & & (S,R;.,.;.,.) & & \\ & \nearrow & & \searrow & \\ (S,S;;,.;.,.) & & & & (L,R;.,.;.,.) \\ & \searrow & & \nearrow & \\ & & (L,S;.,;.,.) & & \end{array}$$

where any combination of L, R and S may be substituted for the first pair ".,." and any combination of A and NA for the second pair ".,.". All these orderings satisfy the conditions P1-P6, and P9 of Kimeldorf and Sampson [130]. The property P7 is satisfied only for intervals containing positive and negative values for the first variable, that is the 32 orderings (L,R;.,.;.,.) and (S,S;.,.;.,.).

The property P8 is satisfied only by the orderings for which the intervals for X and Y can be interchanged, that is (L,R; L,R;.,.) and (L,R; L,R;.,.), (S,R; S,R;.,.) and (S,S;S,S;.,.).

Among these orderings some of them are equivalent : $(S, S; S, S; NA, NA)$ $(S, S; S, S; A, A)$ that is TP2-ordering is the same as P(3,3)-ordering.

Also, $(L, R; L, R; A, A) = (L, R; L, R : NA, NA)$. The ordering on the left of the equality sign was defined by Tchen [216]. It is of interest that the concordance ordering on the left of the equality is equivalent to a more general ordering on the right. This is perhaps an indication of the difficulty inherent in defining fully consistent orderings.

3.7 Bayesian Approach to Stochastic Dependence

Inspired by Bayesian methodology, Brady and Singpurwalla [31] argue that the notion of dependence or independence between two or more variables is conditional on a known or an unknown parameter θ or (latent) variable. For example if (X, Y) is a pair of normal variables, then they are independent or dependent conditionally to their correlation coefficient ρ. Thus if we can define a prior distribution $(\tilde{P})$ on the parameter ρ, we shall be able to associate a certain probability for independence or positive dependence of the pair (X, Y).

Definition 1 :

Suppose that the two variables (X, Y) are independent or quadrant dependent according to the values of a parameter Θ. More precisely :

(1) $F(x, y/\Theta \in I_1) = F_1(x/\Theta \in I_1).F_2(y/\Theta \in I_1), \ \forall(x, y)$
(2) $F(x, y/\Theta \in I_2) \geq F_1(x/\Theta \in I_2).F_2(y/\Theta \in I_2), \ \forall(x, y)$
(3) $F(x, y/\Theta \in I_3) \leq F_1(x/\Theta \in I_3).F_2(y/\Theta \in I_3), \ \forall(x, y)$

where I_i, $i = 1, 2, 3$ is a partition of R. Then the two variables are independent, positive quadrant dependent, or negative quadrant dependent conditionally on Θ in I_1, I_2 or I_3.

Definition 2:

(1) $P((X, Y) \text{ independent}) = \tilde{P}(\Theta \in I_1)$
(2) $P((X, Y) \ PQD) = \tilde{P}(\Theta \in I_2)$
(3) $P((X, Y) \ NQD) = \tilde{P}(\Theta \in I_3)$

If conditions 1, 2 and 3 are fulfilled and the expectations exist, then :

(1) $E(X, Y/\Theta \in I_1) = E(X/\Theta \in I_1).E(Y/\Theta \in I_1)$
(2) $E(X, Y/\Theta \in I_2) \geq E(X/\Theta \in I_2).E(Y/\Theta \in I_2)$
(3) $E(X, Y/\Theta \in I_3) \leq E(X/\Theta \in I_3).E(Y/\Theta \in I_3)$.

Definition 3 :

If ρ is the correlation coefficient between X and Y, and if a prior distribution on ρ is defined, we can compute the probability :

$$\Pi(\alpha) = P(|\rho(X,Y)| \geq \alpha)$$

which is termed by Brady and Singpurwalla a "correlation survival function ".

Definition 4: X and Y are highly stochastically correlated if

$$P(|\rho(X,Y)| \geq \alpha) \geq 1 - \alpha$$

for $0 \leq \alpha \leq 1$.

Definition 5 : The pair (X, Y) is stochastically more correlated than the pair (X', Y') if

$$P(|\rho(X,Y)| \geq \alpha) \geq P(|\rho(X',Y')| \geq \alpha) \ .$$

Definition 6 :

The pair (X, Y) is stochastically more correlated in expectation than the pair (X', Y') if

$$\int \Pi_{X,Y}(\alpha) d\alpha \geq \int \Pi_{X',Y'}(\alpha) d\alpha$$

where $\int \Pi_{X,Y}(\alpha) d\alpha = P(|\rho(X,Y)| \geq \alpha)$.

Orderings of random variables seem to be a fruitful and unexhaustable area of research which attracts both theoretical and applied statisticians, often motivated by newly arising real-world situations.

Chapter 4
Copulas

4.1 Introduction

A copula provides a uniform representation of a bivariate distribution F on the unit square. This result is due to Sklar (1959) [208], and has been revisited by Cartley and Taylor (Section 4.1). Copulas are an important part in the study of dependence between two variables, since they allow us to separate the effect of dependence from effects of the marginals distributions. This feature is analogous to the bivariate normal distribution, where the mean vectors are unlinked from the covariance matrix and jointly determines the distribution. Many authors (Gumbel, Plackett, Mardia, Ali, Mikhail and Haq, Clayton, Joe, Genest, Nelsen – some of them cited in the references) have studied constructions of bivariate distributions with given marginals : this can always be viewed as constructing a copula (Section 4.4). Starting from any real integrable function on the unit square, Rüschendorf (Section 4.5) has proposed a general method for construction of copulas. This method allows us to generate, for example, all the polynomial copulas (Section 4.5.2). Long and Krzysztofowicz utilized a particular case of this method. Another manner to generate copulas is a mixing with respect to a third random variable (called frailty variable). Marshall and Olkin have systematized this idea (Section 4.5.6). Other authors have studied characteristic properties of sub-family of copulas. Genest *et al.* and Sungur *et al.* have considered particularly Archimedean copulas (Section 4.6), Capéraà *et al.* have constructed a more general family of Archimax copulas (Section 4.7). Wei studied copulas with discontinuities (Section 4.8). It is often possible to generalize the method of constructing copulas for more than two variables, but there are compatibility constraints on the marginals. Chakak and Koehler and later Sungur studied a particular case of m–copulas with a property of truncation invariance (Section 4.9.2). A few simulations procedures can be derived from the properties of copulas (Section 4.4.10).

The purpose of this chapter is to present – in an organized manner– numerous results on copulas scattered in diverse literature with the emphasis on dependence concepts and properties. Some results are unified.

4.2 Definition and Some Properties

It is well known and easily verified that $F_1(X)$ and $F_2(Y)$, where F_1 and F_2 are the marginals distributions of X and Y respectively, are two uniform variables if F_1 and F_2 are continuous. Hence if the marginals F_1 and F_2 of the bivariate distribution F are *continuous*, there exists a unique copula, which is a cumulative distribution function, with its marginals being uniform. Formally a function $C : [0,1]^2 \to [0,1]$ such that

$$F(x,y) = C(F_1(x), F_2(y)) \tag{4.1}$$

is a copula. On the other hand, if $C(u_1, u_2)$ and continuous F_1 and F_2 are given, then there exists an F such that:

$$F(F_1^{-1}(u_1), F_2^{-1}(u_2)) = C(u_1, u_2)$$

$F_i(t)$, $i = 1, 2$ is continuous and non decreasing, but could be constant on some intervals. In that case, one defines a quasi-inverse by

$$F_i^{-1}(t) = inf(x : F_i(x) \geq t) \ .$$

Using copulas allows us to separate the study of dependence from the study of the marginals, since one is then reduced to study of the relation between two uniform variables. If the density $f(.,.)$ of $F(.,.)$ exists, one can derive from (4.1), the relationship between the density f of F and c of C:

$$f(x,y) = c(F_1(x), F_2(y)) f_1(x) f_2(y) \ , \tag{4.2}$$

where $f_1(x)$ and $f_2(y)$ are the marginal densities of F.

(1) Elementary properties

$$C(u,1) = P(U \le u, V \le 1) = P(U \le u) = u \tag{4.3}$$

and similarly $C(1,v) = v$. Also

$$C(u,0) = P(U \le u, V \le 0) = 0\,. \tag{4.4}$$

(2) Rectangular inequality
Since $C(u,v)$ is a distribution function, it satisfies for all $0 \le u_1 < u_2 \le 1$ and $0 \le v_1 < v_2 \le 1$

$$\begin{aligned} &Pr(u_1 < U \le u_2, v_1 < V \le v_2) \\ &\quad = C(u_2,v_2) - C(u_1,v_2) - C(u_2,v_1) + C(u_1,v_1) > 0. \end{aligned}$$

This inequality implies in particular that $C(u,v)$ is increasing in both variables. When $C(u,v)$ has a density $c(u,v)$, this inequality becomes $c(u,v) > 0$

(3) Continuity
A copula is continuous in u and v; actually it satisfies the stronger Lipschitz ([196]) condition:

$$|C(u_2,v_2) - C(u_1,v_1)| \le |u_2 - u_1| + |v_2 - v_1|. \tag{4.5}$$

(4) Differentiability
Since $C(u,v)$ is increasing and continuous in the two variables, it is differentiable almost everywhere and from (4.5) we see that $0 \le \frac{\partial}{\partial u}C(u,v) \le 1$ and $0 \le \frac{\partial}{\partial v}C(u,v) \le 1$.

(5) The survival function of a copula
Using (4.3) the survival function $\overline{C}$ corresponding to $C(u,v)$ is :

$$\overline{C}(u,v) = 1 - u - v + C(u,v).$$

From (4.3) and (4.4), we have :

$$\overline{C}(u,1) = 0$$

$$\overline{C}(u,0) = 1\,.$$

Starting from the pair $(1-U, 1-V)$ we can define another copula $C'(u,v)$ whose survival function is connected with C. Namely, given

$$C'(u,v) \quad = \quad Pr(1-U \le u, 1-V \le v) =$$

$$\overline{C}(1-u,1-v) \quad = \quad u+v-1+C(1-u,1-v) \tag{4.6}$$

we then have :

$$\overline{C'}(u,v) = C(1-u,1-v).$$

(6) <u>The dual of a copula</u> The dual of C is the function C_d defined by

$$C_d(u,v) = u + v - C(u,v).$$

Note that C_d is not a copula since it does not satisfy the equation (4.4):$C_d(u,0)$ is equal to u, not zero. However it can easily be verified that $0 \leq C_d(u,v) \leq 1$.

(7) <u>A diagonal copula:</u> A diagonal copula is the single variable function defined by

$$D(u) = C(u,u).$$

We can also define the dual of a diagonal copula $D_d(u) = 2u - D(u)$. If U and V are independent, let C^0 be the corresponding copula. Then $C^0(u,v) = uv$ and $D^0(u) = u^2$.
A diagonal copula can also be viewed as the distribution function of $W = max(U,V)$. Indeed :

$$F_W(w) = Pr(max(U,V) \leq w) =$$
$$Pr(U \leq w, V \leq w) = C(w,w) = D(w).$$

In the same manner, the dual of a diagonal copula $D_d(w')$ is the distribution function of $W' = min(U,V)$.

4.3 The Fréchet Bounds

4.3.1 *Lower and upper Fréchet bounds in the family $\mathcal{F}(F_1, F_2)$ of bivariate distributions with common marginals*

Let $\mathcal{F}(F_1, F_2)$ be the family of the bivariate distributions with common marginals F_1 and F_2. This family has two limits which can be attained. Indeed, since $F(x,y)$ is a non-decreasing function in x and y, then:

$$F(x,y) \leq F(x,\infty) = F_1(x)$$

and

$$F(x,y) \leq F(\infty, y) = F_2(y)$$

i.e.,

$$F(x,y) \leq min(F_1(x), F_2(y)) \,.$$

In the same manner, the survival function S verifies:

$$S(x,y) \leq min(S_1(x)S_2(y))$$

$F^+(x,y) = min(F_1(x), F_2(y))$ is called the upper Fréchet bound.

On the other hand

$$F(x,y) = S(x,y) - 1 + F_1(x) + F_2(y). \tag{4.7}$$

However $S(x,y)$ is always positive or zero, thus:

$$F(x,y) \geq max(0, F_1(x) + F_2(y) - 1)$$

$F^-(x,y) = max(0, F_1(x) + F_2(y) - 1)$ is called the lower Fréchet bound. Similar relations exist between S, S_1 and S_2.

4.3.2 *The Fréchet bounds for a copula*

The supports of the copulas of the Frćhet bounds are the diagonals:the main diagonal to the upper bound, and the other to the lower one. We shall show this result for the upper Fréchet bound $C^+(u,v)$:

$$\begin{aligned} C^+(u,v) &= min(u,v) \\ &= u.1_{[0,v]}(u) + v.1_{]v,1]}(u) \end{aligned} \tag{4.8}$$

where $1_{[0,v]}(u)$ is the indicator function of $[0,v]$. Let v be fixed. The derivative with respect to u (with an extended definition of the derivative, see L. Schwartz [195]) is

$$\begin{aligned} D_u C^+(u,v) &= 1_{[0,v]}(u) + u(-\delta_v(u)) + v(-\delta_v(u)) \\ &= 1_{[0,v]}(u) + (v-u).\delta_v(u) \\ &= 1_{[0,v]}(u) \,, \end{aligned} \tag{4.9}$$

where $\delta_v(u)$ is the Dirac measure on $u = v$. One can write $1_{[0,v]}(u) = 1_{[u,1]}(v)$ and $\delta_v(u) = \delta_u(v)$. Hence the derivative of $g(v) = D_u F^*(u,v)$ with respect to v is

$$D_v D_u C^+(u,v) D_v = \delta_u(v) \, .$$

Therefore, for the upper Fréchet bound of the copula, the measure is concentrated on $\delta_u(v)$, which is the main diagonal ($D_1 : u = v$). Similarly $C^-(u,v) = max(0, u + v - 1)$ has its probability mass concentrated on the second diagonal. In the general case, the upper Fréchet bound is the "curve" ($F_1(x) = F_2(y)$). One can easily verify that the limits of the dual of a copula and of the diagonals copulas are:

$$min(u,v) \leq C_d(u,v) \leq 1 \, ,$$

$$max(0, 2u - 1) \leq D(u) \leq u \, ,$$

and

$$u \leq D_d(u) \leq 1 \, .$$

Using the diagonals of a copula, we can write a condition for the maximal (positive or negative) dependence : If $C = C^+$, then $F_{max(U,V)}(w) = F_{min(U,V)}(w) = D^+(w) = D_d^+(w) = w$, if $C = C^-$, then $F_{max(U,V)}(w) = D^-(w) = max(2w - 1, 0)$ and $F_{min(U,V)}(w) = D_d^-(w) = min(1, 2w)$.

4.4 Examples

(1) The independent copula : $C^0(u,v) = uv$ is the copula associated with a pair (U,V) of independent variables.

(2) A linear convex combination of copulas is a copula. For example:

$$C(u,v) = \alpha . C^+(u,v) + (1 - \alpha) . C^-(u,v) \quad 0 \leq \alpha \leq 1.$$

(3) The family of Farlie-Gumbel-Morgenstern (FGM) copulas are :

$$C_\alpha(u,v) = uv[1 + \alpha(1-u)(1-v)] \quad |\alpha| \leq 1. \tag{4.10}$$

This family does not contain the upper and lower Fréchet bounds.

(4) <u>The Iterated FGM copulas</u>
The survival function of an FGM–copula is

$$\overline{C_\alpha}(u,v) = 1 - u - v + C_\alpha(u,v) = (1-u)(1-v)[1+\alpha.uv]$$

Kotz and Johnson [132] "iterate" the FGM copula by replacing the term $(1-u)(1-v)$ by the survival function $\overline{C_{\alpha_1}}(u,v)$ for some $|\alpha_1| \leq 1$ to obtain :

$$\begin{aligned} C_{\alpha_1,\alpha_2}(u,v) &= uv[1+\alpha_2\overline{C_{\alpha_1}}(u,v)] \\ &= uv[1+\alpha_2(1-u)(1-v)(1+\alpha_1 uv)] \ . \quad (4.11) \end{aligned}$$

Lin used the same idea and "iterated" the FGM copulas by replacing the term uv by $C_{\alpha_1}(u,v)$ in the formula 4.10 and obtained:

$$C_{\alpha_1,\alpha_2}(u,v) =$$
$$uv[1+\alpha_2(1-u)(1-v)(1+\alpha_1.(1-u)(1-v)]. \quad (4.12)$$

See Chapter 5 for additional details of FGM copulas and their generalizations.

(5) <u>Copula associated with the Bivariate Normal Distribution (BVN)</u>
Let (X,Y) be a pair of random variable, with means zero, variances one and bivariate normal distribution: $N((0,0),\Sigma)$ with :

$$\Sigma = \begin{pmatrix} 1 & \rho \\ \rho & 1 \end{pmatrix}.$$

Let ϕ and Φ be respectively the density and the cumulative distribution function of the $N(0,1)$ distribution and ϕ_ρ be the density of the BVN distribution. Let $U = \Phi(X)$ and $V = \Phi(Y)$ be two uniform variables with density c and cumulative distribution function C. Using (4.2), we have:

$$c(u,v) = \frac{\partial^2 C(u,v)}{\partial u \partial v} = \frac{\phi_\rho(\Phi^{-1}(u), \Phi^{-1}(v))}{\phi(\Phi^{-1}(u))\phi(\Phi^{-1}(v))} ,$$

i.e.

$$\begin{aligned} c(u,v) = \frac{1}{\sqrt{1-\rho^2}} . \exp[-\frac{\rho^2}{2(1-\rho^2)}(\Phi^{-1}(u))^2 \\ +(\Phi^{-1}(v))^2 + \rho\Phi^{-1}(u)\Phi^{-1}(v)] \ . \end{aligned}$$

(6) <u>The Plackett copula:</u> Plackett [169] proposed to construct a bivariate distribution for which the ratio

$$\frac{P(U \leq u, V \leq v)P(U > u, V > v)}{P(U > u, V \leq v)P(U \leq u, V > v)}$$

is constant. This idea generalizes to bivariate distributions with continuous marginals the concept of odds-ratio used for testing independence in contingency tables. This ratio can also be written as :

$$\frac{C(u,v).\bar{C}(u,v)}{(u - C(u,v))(v - C(u,v))}.$$

If this ratio is 1, $C = C^0$, i.e. as in contingency tables, the two variables U and V are independent, if the ratio is a constant θ different from one, then $C(u,v)$ is a solution of a quadratic equation:

$$C(u,v) = \frac{1 + (\theta - 1)(u + v) \pm \sqrt{(1 + (\theta - 1)(u + v))^2 - 4uv\theta(\theta - 1}}{2(\theta - 1)}$$

If we require that $0 < \theta$ and $\theta \neq 1$ (i.e. a positive dependence), the solution with the sign $+$ is not appropriate for a copula because in that case $C(u, 0)$ is not zero. The solution with sign $-$ is appropriate ($C(u, 0) = 0$ and $C(u, 1) = u$ and its density is positive). This family includes its upper Fréchet bound : if θ goes to infinity, then C tends to C^+. Hence this family, with only one parameter, is useful for fitting of empirical data.

(7) <u>The Ali-Mikhail-Haq family of distributions</u>

$$C_\theta(u,v) = \frac{uv}{1 - \theta(1-u)(1-v)} \qquad |\theta| \leq 1.$$

If $\theta = 0$, then we have the independence $C_\theta = C^0$. This family does not contain the Fréchet bounds.

4.5 Construction of a Copula

4.5.1 *The Rüschendorf method*

To obtain a general form for a copula, one can use a powerful and elegant method developed by Rüschendorf [186] : let $f^1(u,v)$ has its integral to be zero on the unit square, as well as its two marginals integrate to zero, i.e. :

$$\int_{[0,1]^2} f^1(u,v)dudv = 0 \tag{4.14}$$

and

$$\int_0^1 f^1(u,v)du = 0 \text{ and} \int_0^1 f^1(u,v)dv = 0\,. \tag{4.15}$$

(4.15) implies (4.14). In that case $1 + f^1(u,v)$ is a density of a copula. There is however a constraint, namely that $1 + f^1(u,v)$ must be positive. If it is not the case, but f^1 is bounded, one can then find a constant α such that $1 + \alpha.f^1$ is positive.

It is easy to construct a function f^1 : one starts with an arbitrary real integrable function f on the unit square with its marginals being uniform, and computes:

$$V = \int_{[0,1]^2} f(u,v)dudv\ ,$$

$$f_1(u) = \int_0^1 f(u,v)dv\ ,$$

and

$$f_2(v) = \int_0^1 f(u,v)du\ ,$$

then $f^1 = f - f_1 - f_2 + V$.

If we have two functions f^1 and g^1, possessing the properties stipulated above, then $1 + f^1 + g^1$ is the density of a copula, and more generally $1 + \Sigma_{i=1}^n f_i^1$ is a density with f_i^1 satisfying (4.14) and (4.15).

4.5.2 *Application to polynomial copulas*

To obtain a polynomial copula one starts with $f = u^k v^q$ and obtains

$$f^1(u,v) = (u^k - \frac{1}{k+1})(v^q - \frac{1}{q+1}) \ , k \geq 1 \ , q \geq 1 \ .$$

Therefore the function :

$$1 + \theta(u^k - \frac{1}{k+1})(v^q - \frac{1}{q+1}) \tag{4.16}$$

with the constraint :

$$0 < \theta \leq min\left(\frac{(k+1)(q+1)}{q}, \frac{(k+1)(q+1)}{k}\right)$$

is the density of a polynomial copula. Combining the expressions (4.16) for all k, $k \geq 1$, and all q, $q \geq 1$, we arrive at the general formula:

$$\frac{\partial^2 C}{\partial u \partial v} = 1 + \sum_{k\geq 1, q \geq 1} \theta_{kq}(u^k - \frac{1}{k+1})(v^q - \frac{1}{q+1}) \tag{4.17}$$

with the constraints :

$$0 \leq min(\sum_{k\geq 1; q\geq 1} \theta_{kq} . \frac{q}{(k+1)(q+1)}, \sum_{k\geq 1; q\geq 1} \theta_{kq} . \frac{k}{(q+1)(k+1)}) \leq 1$$

and, after some simplifications, a polynomial copula of power m can be written as

$$C(u,v) = uv[1 + \sum_{k\geq 1, q\geq 1}^{k+q \leq m-2} \frac{\theta_{kq}}{(k+1)(q+1)}(u^k - 1)(v^q - 1)]. \tag{4.18}$$

If one requires, for example, a polynomial copula of the fifth power, its density is written as :

$$\frac{\partial^2 C}{\partial u \partial v} = 1 + \theta_{11}(u - \frac{1}{2})(v - \frac{1}{2}) + \theta_{12}(u - \frac{1}{2})(v^2 - \frac{1}{3}) + \theta_{21}(u^2 - \frac{1}{3})(v - \frac{1}{2}) \ .$$

The polynomial copula of the fifth power then becomes:

$$C(u,v) = uv[1 + \frac{\theta_{11}}{4}(u-1)(v-1) + \frac{\theta_{12}}{6}(u-1)(v^2-1) + \frac{\theta_{21}}{6}(u^2-1)(v-1)] \ . \tag{4.19}$$

Using this method, we can generate all polynomial copulas. Indeed the dimension of the space of polynomial copulas of power m is equal to the number of the parameters θ_{kq}: k varies from 1 to $(m-3)$, the range of q is the same, but $q+k \leq m-2$, thus number of parameters is then $\sum_{i=1}^{m-3} i = \frac{(m-3)(m-2)}{2}$. This dimension coincides with the one found by Wei *et al.* [223], using a different argument for polynomial copulas of power m.

Expanding the expression (4.19), we can verify that it coincides with the expression furnished by Wei *et al.* [223] using a different approach.

4.5.2.1 *Approximation of a copula by a polynomial copula*

If a copula, indexed by a parameter θ, $C_\theta(u,v)$ belongs to the C^n functions with respect to θ (the functions whose n-th derivative is continuous), we can then express $C_\theta(u,v)$ by means of the Taylor expansion in the neighborhood of θ_0:

$$C_\theta(u,v) \approx C_{\theta_0}(u,v) + \sum_{i=1}^{i=n} \frac{C_{\theta_0}^{(i)}(u,v)(\theta-\theta_0)^i}{i!}.$$

Choosing the value θ_0 which yields the independence i.e. such that $C_{\theta_0}(u,v) = uv$, and if the successive derivatives of C_θ with respect to θ are powers in uv, we then obtain an approximation of C_θ by means of a polynomial copula.

Examples:

(1) The FGM family:
The FGM family corresponds to its first order expansion in Taylor's series around $\theta = 0$

(2) The Ali-Mikhail-Haq family :

$$C_\theta(u,v) = \frac{uv}{1-\theta(1-u)1-v)} = uv[1+\sum_{i\geq 1}(\theta(1-u)(1-v))^i] \qquad |\theta| \leq 1. \tag{4.20}$$

If we consider in (4.20) the first-order approximation only, we then have the FGM family, and with the second order approximation we arrive at the iterated FGM family of Lin. For an approximation of any order we have a polynomial copula.

(3) The Plackett family: Nelsen ([161], page 83) proved that FGM family is a first order approximation to the Plackett family, by expanding it in Taylor's series around $\theta = 1$.

(4) The copula associated to BVN family: The first order approximation around $\rho = 0$ is

$$C_\rho(u,v) = uv[1 + \rho\Phi^{-1}(u)\Phi^{-1}(v)] .$$

4.5.3 *Other examples*

Rüschendorf [186] gives two other examples of constructing copulas :

(1) A distribution concentrated in the vicinity of the diagonal :
Let

$$f(u,v) = \frac{1}{\sqrt{|u-v|}} .$$

The function $f(u,v)$ is not a density of a copula, but one can compute $f^1 = f - f_1 - f_2 + V$ where

$$f_1(u) = 2\sqrt{u} + 2\sqrt{1-u}$$

with $f_2(v)$ obtained by interchanging u and v, and

$$V = \int_0^1 f_1(u)du = \frac{8}{3} .$$

Hence:

$$1 + f^1(u,v) = \frac{1}{\sqrt{|u-v|}} - 2\sqrt{u} - 2\sqrt{1-u} - 2\sqrt{v} - 2\sqrt{1-v} + \frac{8}{3}$$

is a density of a copula.

(2) A not concentrated distribution:
Let

$$f_\theta(u,v) = -1_{|u-v|>\theta} \quad 0 < \theta < 1$$

where $1_{|u-v|>\theta} = 1$ when $|u-v| > \theta$ and 0 otherwise. Then the first marginal is

$$f_{1\theta}(u) = (1-2\theta)1_{\theta \le u \le 1-\theta} + (1-u-\theta)1_{0 \le u < \theta} - (u-\theta)1_{u \ge 1-\theta}$$

for $0 \le \theta \le 0.5$ and

$$f_{1\theta}(u) = (u-\theta)1_{u>\theta} + (\theta-u)1_{u<1-\theta} \quad \textit{for } 0.5 \le \theta \le 1.$$

Interchanging u by v in the above expression, we have a formula for the other marginal $f_{2\theta}$. Finally the associated density is

$$1 + f_\theta^1(u,v) = 1 + f_\theta(u,v) - f_{1\theta}(u) - f_{2\theta}(v) + \theta(1-\theta).$$

If $\theta = 1$, we have the independence case.
Our third example is the one proposed by Long and Krzysztofowicz [146], who may not have been aware of the work of Rüschendorf.

(3) Copula with scalar parameter and a power function characteristic: Long and Krzysztofowicz [146] express the density function of a copula as

$$f(u,v) = 1 + \theta c(u,v)$$

with the same constraints on θ (a scaling parameter) and $c(u,v)$ as Rüschendorf. Next using the ideas of Kotz and Seeger [132] related to the methods of constructing densities of a copula (termed in the paper of Kotz and Seeger "a density weighting function"), they specify the function $c(u,v)$ in such a manner that it possesses symmetries about the main and second diagonals, and in case of positive dependence it has a ridge along the main diagonal (the upper Fréchet bound) and a through along the second diagonal (the lower Fréchet bound). The structure is opposite in the case of negative dependence.
More precisely, they utilize a function κ, called the regression characteristic to define $c(u,v)$. κ is defined on $[0,1]$ by $\kappa(t) = (1-t)^\beta$ or $\kappa(t) = t^\beta$ for $\beta > 0$. Now :

$$c(u,v) = c_1(u,v) + c_2(u,v) - 2K(1)$$

where the function $K(u) = \int_0^u \kappa(t)dt$, and

$$c_1(u,v) = \kappa(|v-u|)$$

$$c_2(u,v) = \begin{cases} \kappa(u+v) & \text{for } u \leq 1-v \\ \kappa(2-u-v) & \text{for } u > v. \end{cases}$$

If $\kappa(t)$ is decreasing and $\theta > 0$, then the pair (U,V) has a positive dependence with a ridge along the main diagonal due to $c_1(u,v)$ and a through along the second diagonal due to $c_2(u,v)$, and a negative dependence if $\theta < 0$. Conversely, if $\kappa(t)$ is increasing and

$\theta > 0$, then the pair has negative dependence with a ridge along the second diagonal and a through along the main one.
Parameters θ and β are interconnected. Namely $K(1) = \frac{1}{\beta+1}$ and $0 \leq \kappa(t) \leq 1$, imply that $\frac{-2}{\beta+1} \leq c(u,v) \leq 1$ and since the density $1 + \theta c(u,v)$ is positive we have :

$$-\frac{\beta+1}{2\beta} \leq \theta \leq \frac{\beta+1}{2}.$$

For this model the correlation coefficient of (U,V) depends on β slightly, but β affects substantially the conditional variance $var(V/U = u)$. If $\kappa(t) = (1-t)^{\beta}$, the correlation coefficient is between $-7/12$ and 1 for $\beta < 1$, and between 0 and 1 for $\beta > 1$. On the other hand, if $\kappa(t) = t^{\beta}$, then the correlation coefficient is between -1 and $7/12$. Therefore it is possible to obtain all the degrees of dependence between U and V using this model.

4.5.4 *Models defined from a distortion function*

In the field of insurance pricing (see e.g. Frees and Valdez [75]) one uses a distortion function g which satisfies $g : [0,1] \mapsto [0,1]$, with $g(0) = 0$, $g(1) = 1$ and g is increasing. Starting with the distribution function $F(x,y) = C[F_1(x), F_2(y)]$, one defines another distribution function $\tilde{F}(x,y) = g[F(x,y)]$ with marginals $g(F_1)$ and $g(F_2)$. The associated copula is then $\tilde{C}(u,v) = g[C(g^{-1}(u), g^{-1}(v))]$. For example with an exponential distortion function:

$$g(t) = \frac{1-e^{\alpha.t}}{1-e^{-\alpha}},\ \alpha > 0$$

and with an independent copula $C(u,v) = uv$, we find :

$$\tilde{C}(u,v) = \log_{\alpha}(1 + \frac{(\alpha^u - 1)(\alpha^v - 1)}{\alpha - 1})$$

which is the well known Frank's copula (Frank, 1979 [72]).

4.5.5 *Frailty models*

The frailty models have been defined and widely used in the field of survival analysis (see, for example, Kalbfleisch and Prentice [122] page 33, Hougaard [100], Oakes [163]). Recall that $h(x) = -\frac{d}{dx} lnS(x)$ is the marginal hazard of X, $H(x) = \int_0^x h(u)du = -lnS(x)$ the cumulative hazard of X. In these models, a frailty variable W, with distribution function $G(w)$ acts multiplicatively on the marginal hazards. For example for the random variable X:

$$h_1(x/w) = h_{10}(x).w,$$

where $h_{10}(x)$ is the baseline hazard. Noting that $S_{10}(x) = exp(-H_{10}(x))$ we obtain

$$S_1(x/w) = exp(-\int_0^x h_{10}(u)du.w) = exp(-H_{10}(x)w) = (S_{10}(x))^w .$$

Therefore the marginal $S_1(x)$ is

$$S_1(x) = \int exp(-H_{10}(x)w).dG(w) = \int S_{10}^w(x).dG(w).$$

Furthermore, one supposes that, conditionally to W, the variables X and Y are independent. Hence, we have :

$$S((x,y)/w) = S_1(x/w).S_2(y/w) = exp[-w(H_{10}(x) + H_{20}(y))] ,$$

and integrating:

$$S(x,y) = \int \exp(-w[H_{10}(x) + H_{20}(y)])dG(w) . \tag{4.21}$$

Finally

$$S(x,y) = E(\exp(-wH_0))$$

with $H_0(x,y) = H_{10}(x) + H_{20}(y)$. Thus $S(x,y)$ is defined on $]0,\infty[^2$ with the Laplace transform φ^{-1} (the reason for this designation will be clarified in the next section). But, since $S_1 = \varphi^{-1}(H_{10})$, we have :

$$S(x,y) = \varphi^{-1}[\varphi(S_1(x)) + \varphi(S_2(y))] .$$

In the case of copulas, these models are a particular case of the popular archimedian copulas (see the next section).

4.5.6 *Marshall and Olkin's generalization*

In frailty models, the survival function S is obtained from the marginal survival functions S_i, $i = 1, 2$ by Laplace transform. This procedure can also be applied to F_i, $i = 1, 2$ to obtain another distribution function F :

$$F(x, y) = \varphi^{-1}[\varphi(F_1(x)) + \varphi(F_2(y))] .$$

More generally, let $G(w_1, w_2)$ be a mixture distribution defined on $]0, \infty[^2$ with the Laplace transform φ^{-1}, and its marginals G_i ,$i = 1, 2$ with the Laplace transforms φ_i^{-1}, K a bivariate distribution function with uniform marginals on $[0, 1]$. F_1 and F_2 are defined using F_{10} and F_{20} the two univariate baseline distributions ($F_i = \varphi_i(-lnF_{i0})$, $i = 1, 2$) then there exists a distribution function F such that [152]

$$F(x_1, x_2) = \int\int K(F_{10}^{w_1}(x_1), F_{20}^{w_2})(x_2)dG(w_1, w_2). \tag{4.22}$$

A straightforward generalization for the case of n ($n \geq 2$) variables yields :

$$F(x_1, .., x_n) = \int\int K(F_{10}^{w_1}(x_1), ..., F_{n0}^{w_n}(x_n))dG(w_1, ..., w_n). \tag{4.23}$$

Various G and K functions result in a variety of multivariate distributions functions (or survival functions) defined by means of their marginals. For example:

(1) K is the independent bivariate distribution and G is the upper Fréchet bound (UFB):
In that case $G(w_1, w_2) = min_i(G_i(w_i))$, and we retrieve the frailty model defined in the preceding section:

$$F(x_1, x_2) = \varphi^{-1}[\varphi(F_1(x_1)) + \varphi(F_2(x_2))]$$

which can be generalized to case of n ($n \geq 2$) variables :

$$F(x_1, ..., x_n) = \varphi^{-1}[\Sigma_{i=1}^n \varphi(F_i(x_i))] . \tag{4.24}$$

Various examples with the Gamma and stable Laplace tansforms are provided in the next section (Archimedean copulas). In that case, the function $F(x_1, ..., x_n)$ is TP_∞ (totally positive of order infinity in each pair of arguments), and in particular it is TP_2 (see Chapter 3, section 3.2.8).

If G has a negative binomial distribution with the Laplace transform $\varphi^{-1}(t) = \left(\frac{pe^{-t}}{1-qe^{-t}}\right)^{\alpha}$, $\alpha > 0$, $0 < p < 1$, $q = 1 - p$, and the inverse function $\varphi(t) = -ln(\frac{t^{\frac{1}{\alpha}}}{p+qt^{\frac{1}{\alpha}}})$ then

$$F(x_1, x_2) = \frac{F_1(x_1)F_2(x_2)}{[1 - q(1 - F_1^{\frac{1}{\alpha}}(x_1))(1 - F_2^{\frac{1}{\alpha}}(x_2))]^{\alpha}}. \tag{4.25}$$

The Ali–Mikhail–Haq distribution is a particular case where $\alpha = 1$, with a more relaxed definition of $-1 \leq q \leq 1$.

(2) K is an independent bivariate distribution, G has convolution form: We shall assume here that the frailty variables are $W_1 = U + W$ and $W_2 = V + W$, where U, V, W are independent. Then the Laplace transform of the pair (W_1, W_2) is $\varphi^{-1}(t_1, t_2) = \psi_1(t_1)\psi_2(t_2)\psi_0(t_1 + t_2)$ where $\psi_1(t_1)$, $\psi_2(t_2)$ and $\psi_0(t)$ are the Laplace transforms of U, V and W respectively, and the univariate Laplace transforms are

$$\varphi_1^{-1}(t) = \varphi^{-1}(t, 0) = \psi_1(t)\psi_0(t) ,$$

$$\varphi_2^{-1}(t) = \varphi^{-1}(t, 0) = \psi_2(t)\psi_0(t) .$$

In this case

$$F(x_1, x_2) = \psi_1[\varphi_1 F_1(x_1)].\psi_2[q_2 F_2(x_2)].\psi_0[\varphi_1 F_1(x_1) + \varphi_2 F_2(x_2)].$$

Specifically, If

$$\psi_i(t) = \left(\frac{p.e^{-t}}{1 - qe^{-t}}\right)^{\alpha_i} \quad \alpha_i \geq 0, \ i = 1, 2, 3 ,$$

$\alpha_{10} = \alpha_1 + \alpha_0 > 0$,$\alpha_{20} = \alpha_2 + \alpha_0 > 0$, then

$$q_1(t) = -ln(\frac{t^{1/\alpha_{10}}}{p + qt^{1/\alpha_{10}}})$$

$$q_2(t) = -ln(\frac{t^{1/\alpha_{10}}}{p + qt^{1/\alpha_{10}}})$$

and

$$F(x_1, x_2) = \frac{F_1(x_1).F_2(x_2)}{[1 - q(1 - F_1^{\frac{1}{\alpha_{10}}}(x_1)(1 - F_2^{\frac{1}{\alpha_{20}}}(x_2))^{\frac{1}{\alpha_{20}}})]^{\alpha_0}}.$$

This is a generalization of the family (4.25) obtained with a negative binomial Laplace transform.
Another application is the correlated gamma frailty model of Yashin and Iachine [230], written for the survival function and where U, V, and W have gamma distributions, U and V with mean $1-\rho$ and variance θ, and W with mean ρ and variance $\rho\theta$. The resulting variables W_1 and W_2 have gamma distribution with mean 1 and variance θ, and correlation coefficient ρ. The joint survival function is

$$S(x_1, x_2) = S_1(x_1)^{1-\rho} S_2(x_2)^{1-\rho}[S_1(x_1)^{-\theta} + S_2(x_2)^{-\theta} - 1]^{-\rho/\theta}.$$

In case of n variables, a straightforward generalization is:let U_i, $i = 1, ..., n$ be a family of gamma variables with mean $1-\rho$ and the same variance θ, and W a gamma variable with mean ρ and variance $\rho\theta$. Then, $W_i = U_i + W \sim \Gamma(\frac{\rho}{\theta}, \frac{1}{\theta})$, $i = 1, ..., n$ and :

$$S(x_1, ..., x_n) = \Pi_{i=1}^n S_i(x_i)^{1-\rho}[\Sigma_{i=1}^n S_i(x_i)^{-\theta} - n + 1]^{-\rho/\theta}.$$

Yaashin and Iachine [230] give a more general formula where the variances of the U_i, $i = 1, ..., n$ can be different.

(3) K is the UFB, G is an independent bivariate distribution with exponential marginals: Explicitly

$$K(u, v) = min(u, v) \ \ 0 \leq u, v \leq 1$$

and

$$F(u, v) = \int\int min(u^{w_1}, v^{w_2}).w_1 e^{-w_1}.w_2 e^{-w_2}.dw_1 dw_2.$$

In this case the marginals are:

$$\begin{aligned} F_1(u) &= \int_0^\infty u^{w_1}.w_1 e^{-w_1}.dw_1 \\ &= \int_0^\infty w_1 e^{-w_1(1-log u)} dw_1 \end{aligned}$$

$$= \frac{1}{1 - logu} \tag{4.26}$$

and $F_2(v) = \frac{1}{1-logv}$.
After a number of transformations we obtain :

$$F(u,v) = F_1(u)F_2(v)(1 + \frac{(1-F_1(u))(1-F_2(v))}{1-F_1(u)F_2(v)}).$$

This expression involves no parameter, but its convex combination with the independence case gives :

$$\begin{aligned} F'(u,v) &= (1-\alpha)F_1(u)F_2(v) + \alpha F(u,v) \\ &= F_1(u)F_2(v)\left(\frac{\alpha S_1(u)S_2(v)}{1-F_1(u)F_2(v)}\right) \; 0 < \alpha < 1 \, . \end{aligned}$$

(4) K is the lower Fréchet bound, G is an independent bivariate distribution:
Explicitly

$$K(u,v) = max(u+v-1,0), \; 0 \leq u,v \leq 1 \, .$$

Marshall and Olkin [152] provide an example where G possesses independent exponential marginals. They arrive at the distribution:

$$F(u,v) = F_1(u_1) + F_2(u_2) - 1 + B(\frac{1}{S_1(u_1)}, \frac{1}{S_2(u_2)})$$

where $B(.,.)$ is the beta function.

This method of using mixtures of distributions has been developed by Joe and Hu [111] for the case of max-infinitely divisible distributions (i.e distributions for which all the positive powers of it are also proper distributions functions). A particular case of these distribution are the extreme value distributions for maxima. Joe and Hu [111] generate many multivariate (not only bivariate) families of distributions functions with positive dependence and study properties of closure properties in these family as well as concepts of dependence including the tail dependence.

4.6 Archimedean Copulas

4.6.1 *Definition and basic properties*

An important particular case of copulas are Archimedean copulas (we note again that frailty models are their particular cases). Genest, Nelsen and Sungur among others, studied these copulas in great detail [81], [83], [160], [212]. If φ, called a generator, is a convex decreasing function, with a positive second derivative, and $\varphi(1) = 0$ and such that

$$\varphi : (0,1] \to [0, +\infty].$$

One can then define an inverse (or quasi-inverse if $\varphi(0) < \infty$) by :

$$\varphi^{-1}(x) = \begin{cases} \varphi^{-1}(x) & \text{for } 0 \leq x \leq \varphi(0) \\ 0 & \text{for } \varphi(0) < x < +\infty\,. \end{cases}$$

An Archimedean copula is then defined as:

$$C(u,v) = \varphi^{-1}[\varphi(u) + \varphi(v)]. \tag{4.27}$$

Remark 1: The existence of the second derivative φ'' allows to calculate a density $c(u,v)$ when $\varphi(u) = \varphi(v) < \varphi(0)$. Indeed in that case, if D_i $i = 1,2$ are the operators for the partial derivatives with respect to u and v, $\varphi(C(u,v)) = \varphi(u) + \varphi(v)$ implies

$$\varphi'(C)D_1C = \varphi'(u)\,, \tag{4.28}$$

$$\varphi'(C)D_2C = \varphi'(v)\,. \tag{4.29}$$

Therefore $D_1(C) = \frac{\varphi'(u)}{\varphi'(C)}$, and differentiating with respect to the second variable we obtain an explicit expression for D_1D_2C :

$$\begin{aligned} D_1D_2C &= -\frac{\varphi'(u)\varphi''(C)D_2C}{\varphi'(C)^2} \\ &= -\frac{\varphi'(u)\varphi''(C)\varphi'(v)}{\varphi'(C)^3}\,. \end{aligned} \tag{4.30}$$

Remark 2: The convexity and the decreasing property of φ (which implies convexity and the decreasing property of φ^{-1}) are necessary and sufficient to assure that the density $c(u,v)$ is greater than zero. This is easy to verify using Eq. (4.30), and the fact that the convexity is equivalent to an increasing first derivative, i.e. positivity of the second derivative.

Remark 3: With the same generator φ, one can define a complementary copula by using the survival functions S, S_1 and S_2 in place of C, U and V. Indeed starting from Eq. (4.6):

$$C'(u,v) = u + v - 1 + C(1-u, 1-v)$$

the survival function of C' is

$$S(u,v) = \bar{C}'(u,v) = C(1-u, 1-v).$$

Hence one can define

$$S(u,v) = \varphi^{-1}[\varphi(1-u) + \varphi(1-v)]\,.$$

i.e.

$$S(u,v) = \varphi^{-1}[\varphi(S_1(u)) + \varphi(S_2(v))]\,.$$

Remark 4: The generator φ is not unique, in fact $c\varphi$, where c is a positive constant yields the same copula as φ.

Remark 5: If $\varphi(0) \neq \infty$, i.e. if φ^{-1} is a quasi-inverse, so that the copula possesses a singular part, which is concentrated on a set of measure zero (with respect to the Lebesgue measure), which corresponds to the curve :

$$\{(u,v)\ :\ \varphi(u) + \varphi(v) = \varphi(0)\}.$$

Genest [81] has shown that the probability of this set is $-\varphi(0)/\varphi'(0)$. Indeed

$$\int\!\!\int_{\varphi(u)+\varphi(v)<\varphi(0)} c(u,v)dudv = -\int_0^1 \varphi'(t)dt \int_0^t \frac{\varphi''(z)dz}{\varphi'(z)^2}$$

$$= \int_0^1 \varphi'(t)dt[\frac{1}{\varphi'(t})-\frac{1}{\varphi'(0)}] = 1 + \frac{\varphi(0)}{\varphi'(0)} \qquad (4.31)$$

where $c(u,v)$ is the density of the copula, and using Eq. (4.30), with the change of variables $z = C(u,v)$ and $t = u$ we verify the assertion.

4.6.2 *Examples*

(1) Let the generator be $\varphi(t) = (1-t)^\alpha\ \ \alpha \geq 1$

In this case :

$$C(u,v) = max(0, 1 - [(1-u)^\alpha + (1-v)^\alpha]^{1/\alpha}).$$

The lower Fréchet bound is attained when α tends to one, and the upper when α tends to infinity.

(2) The Clayton family [46] $\varphi(t) = (1/t)^{\alpha-1} - 1, \quad \alpha > 1$. This family (also studied by Cook and Johnson [48]) is constructed from a frailty variable W having a Gamma distribution with the Laplace transform $\varphi^{-1}(t) = (1+t)^{1/1-\alpha}$, $\alpha > 1$.
In this case :

$$C_\alpha(u,v) = (u^{1-\alpha} + v^{1-\alpha} - 1)^{\frac{1}{1-\alpha}} \qquad \alpha > 1.$$

As the parameter α tends to infinity, dependence becomes maximal, and as α tends to one, the pair (U,V) becomes independent. It is possible to extend this family to have negative dependence, using the generator $\varphi(t) = 1 - t^{1-\alpha}$ $0 < \alpha < 1$. In that case $C(u,v) = max(u^{1-\alpha} + v^{1-\alpha} - 1, 0)^{1/1-\alpha}$. As $\alpha \to 0$ the distribution tends to the lower Fréchet bound. Each member of this family has LRD dependence as it is shown Chapter 3, Section 3.2.8. But the family is not ordered by the DTP(0,1)-ordering, see Chapter 3, Section 3.6.6.

(3) The Frank family [72],[82]
In this case

$$\varphi(t) = \ln \frac{(1-\alpha)}{(1-\alpha^t)} \; , \; 0 < \alpha < 1 \, .$$

This family is also obtained by a Laplace transform of a logarithmic series on positive integers. We have

$$C(u,v) = \log_\alpha \left(1 + \frac{(\alpha^u - 1)(\alpha^v - 1)}{(\alpha - 1)}\right).$$

The dependence becomes maximal when α tends to zero and independence is achieved when $\alpha = 1$. It is possible to extend this family to the case of negative dependence with $\alpha > 1$ (see Meester and MacKay [154]). In this situation as α tends to infinity the copula tends to its lower Fréchet bound.

(4) The Gumbel-Hougaard family [88] [99] (also known as bivariate logis extreme value distribution).

$$\varphi(t) = (-\ln(t))^{1/\alpha}, 0 < \alpha < 1.$$

This family is obtained by the Laplace transform of a positive stable distribution ($\varphi^{-1}(t) = exp(-t^{\alpha})$). It plays a prominent role among extreme value distributions. It is the only Archimedean copula which is also an extreme value copula. We have:

$$C(u, v) = \exp\left(-[(-\ln(u))^{1/\alpha} + (-\ln(v))^{1/\alpha}]^{\alpha}\right)$$

As α tends to zero, the dependence becomes maximal, and as α tends to zero the pair (U, V) becomes independent. No extension to negative dependence is available.

(5) The Ali-Mikhail-Haq family [4]: Setting

$$\varphi_{\alpha}(t) = (1-\alpha)^{-1} \log[\frac{1+\alpha(t-1)}{t}] \qquad , |\alpha| \leq 1$$

we retrieve the Ali-Mikhail-Haq family (already mentioned in Section 4.5.6). This family does not attain the Fréchet bounds. This distribution as also the three preceding ones, which come from a Laplace transformation verify the strong property of LRD-dependence (Chapter 3 Section 3.2.6).

(6) The lower Fréchet bound of a copula is Archimedean. Specifically $C^{-}(u, v) = max(0, u + v - 1)$ is generated by $\varphi(t) = 1 - t$, $0 \leq t \leq 1$.

(7) The independent copula: With the generator $\varphi(t) = -ln(t)$, we generate the independent copula. Namely:

$$C^0(u, v) = u.v = exp[-(-ln(u)) + (-ln(v))].$$

(8) The Upper Fréchet Bound is not Archimedean.

(9) The FGM family: We will show in the next Section, that this family is not Archimedean.

4.6.3 *A characterization of Archimedean copulas*

(1) The Abel (1826) [2] criterion
The theorem below, due to Genest [81] provides a characterization of Archimedean copulas. Its proof involves an identity related to

symmetric bivariate functions proved by Abel in 1826. For this reason Genest refers to his theorem as the Abel criterion for copulas.
Theorem : A copula C is Archimedean if (and only if), there exists a mapping $f : (0,1) \to (0,\infty)$ such that :

$$\frac{\frac{\partial C(u,v)}{\partial u}}{\frac{\partial C(u,v)}{\partial v}} = \frac{f(u)}{f(v)} \qquad \forall(u,v)\;;0 < u,v < 1.$$

The function φ is given (up to a constant) by :

$$\varphi(t) = \int_t^1 f(u)du\;.$$

The necessity of the criterion is easy to prove. From Eqs. (4.28) and (4.29), we deduce

$$\frac{\frac{\partial C}{\partial u}}{\frac{\partial C}{\partial v}} = \frac{\varphi'(u)}{\varphi'(v)}\;.$$

To show the converse, Genest utilizes Abel criterion. [2]
We can now verify that FGM family is not Archimedean. Indeed in this case :

$$\frac{\frac{\partial C(u,v)}{\partial u}}{\frac{\partial C(u,v)}{\partial v}} = \frac{v(1+\theta(1-v)(1-2u)}{u(1+\theta(1-u)(1-2v)}$$

which is not of the form $\frac{f(u)}{f(v)}$, except when U and V are independent (i.e $\theta = 0$).

(2) The Ling (1965) theorem
One of the earliest results on Archimedean copulas is Ling's theorem which states that Archimedean copulas are the only copulas which satisfy the two conditions :

(a) associativity : $\forall(u,v)C(C(u,v),w) = C(u,C(v,w))$
(b) $\forall u \in]0,1[\;, C(u,u) < u$.

4.6.4 *The limit of a sequence of Archimedean copulas*

Let C_n ,$n \in N$ $n \geq 1$ be a family of Archimedean copulas, and let φ_n, the family of generators. Under what conditions, is the limit (if exists) $C = \lim_{n\to\infty} C_n$ an Archimedean copula? Denote by $\mathbf{\Phi}$, the set of functions, which are continuous, strictly decreasing and convex, with a continuous second derivative. The functions φ_n belong to Φ.

Proposition 1 (Genest): Assume that the copulas C_n are absolutely continuous with respect to the Lebesgue measure. The limit C is an Archimedean copula if and only if, there exists $\varphi \in \Phi$, such that for all s and all t in $[0,1]$:

$$lim_{n\to\infty}\frac{\varphi_n(s)}{\varphi_n'(t)} = \frac{\varphi(s)}{\varphi'(t)}. \tag{4.32}$$

If Eq. (4.32) is valid, then there exists a sequence of constant c_n, such that $lim_{n\to\infty}c_n\varphi_n(t) = \varphi(t)$

Proposition 2 (Genest): Let φ_n, $n \geq 1$, be a sequence of functions belonging to Φ , and let C_n be the sequence of associated copulas. Then:

$$\forall x\ \forall y\ \in [0,1]\ \lim_{n\to\infty} C_n(u,v) = min(u,v)$$

if and only if $lim_{n\to\infty}\frac{\varphi_n(t)}{\varphi_n'(t)} = 0$ for all t in $[0,1]$.

Example: Choosing $\varphi_n(t) = t^{-\frac{1}{n}} - 1$, and $c_n = n$, we have :

$$lim_{n\to\infty}n\varphi_n(t) = -logt\ . \tag{4.33}$$

Hence:

$$lim_{n\to\infty}C_n(u,v) = uv\ .$$

The sequence $C_n(u,v)$ is a subfamily of the Clayton family mentioned in the previous Section.

4.6.5 *Characterization of Archimedean copulas by their diagonal copulas*

Equation

$$\varphi(C(u,v)) = \varphi(u) + \varphi(v)$$

applied to the diagonal copula yields

$$\varphi(D(u)) = 2\varphi(u)$$

for all $u \in [0,\varphi(0)]$ i.e. $D(u) = \varphi^{-1}(2\varphi(u))$.

Conversely, given D, the diagonal of an Archimedean copula, is it possible to define uniquely a generator φ, such that $\varphi(D(u)) = 2\varphi(u)$? Moreover can we retrieve an Archimedean copula with this generator? This problem was studied by Sungur and Yang [212]. The answer is yes provided D satisfies the following conditions:

(1) $D : [0,1] \to [0,1], \quad D(0) = 0, \quad D(1) = 1$;
(2) $D(u)$ is a strictly increasing function in u ;
(3) $D(u) < u \quad if\ u \in]0,1[$;
(4) $\forall u \in]0,1[, \quad \lim_{n\to\infty} D^n(u) = 0 \quad , \lim_{n\to\infty} D^{-n}(u) = 1$ where $D^n = D \circ D \circ \ldots \circ D$ and $D^{-n} = D^{-1} \circ D^{-1} \circ \ldots \circ D^{-1}$
(5) $\forall u \in (0,1), \quad \lim_{n\to\infty} 2^n(D^{-n}(u) - 1) < \infty$.

The two first conditions are fulfilled if D is a diagonal copula. The strictly increasing property of $D(u)$ in u is a direct consequence of the strict decreasing of $\varphi(u)$ in u. The next condition stems from the Ling theorem (Section 4.6.3), since D is the diagonal of an Archimedean copula. The fourth condition is easily derived from conditions 1 and 3. Only the fifth condition is directly connected with the construction of φ.

- Proof of the necessity of the conditions: From the fourth condition, one can deduce that :

$$\lim_{n\to\infty} \frac{\varphi(D^{-n}(u) - \varphi(1)}{D^{-n}(u) - 1} = \varphi'(1).$$

However $\varphi(1) = 0$ and $\varphi(D^{-n}(u) = 2^{-n}\varphi(u)$, hence :

$$\lim_{n\to\infty} \frac{2^{-n}\varphi(u)}{D^{-n}(u) - 1} = \varphi'(1)$$

Therefore if $\lim_{n\to\infty} 2^n(D^{-n}(u) - 1)$ exists, we arrive at

$$\forall u \in]0,1[\ \ \varphi(u) = \varphi'(1). \lim_{n\to\infty} 2^n(D^{-n}(u) - 1) .$$

We know that $\varphi(u)$ is defined up to a constant, that is $c\varphi(u)$, where c is a positive constant, generates the same copula. Hence utilizing the fact that $\varphi'(1)$ is always negative, we have the relationship :

$$\varphi(u) = - \lim_{n\to\infty} 2^n(D^{-n}(u) - 1) . \tag{4.34}$$

- We now sketch a proof of the sufficiency condition. With φ defined as in the Equation (4.34), is φ the generator of an Archimedean copula ? Essentially, we have to show that φ is convex, decreasing, and that $\varphi(1) = 0$.

Let the sequence be $\varphi_n(u) = (1 - D^{-n}(u))$ and the constant be $c_n = 2^n$. We have: $\varphi_n(1) = 0$ for all n and $\varphi_n(0) = 1$. Thus $\varphi(1) = 0$ and $\varphi(0) = lim_{n\to\infty} 2^n = \infty$.

- Since $D(u) < u$ for all u in $]0,1[$ and $D(u)$ strictly increasing , $D(0) = 0$ and $D(1) = 1$ thus D is convex, however since D is strictly increasing, D^{-1} is concave and so is D^{-n}. Hence $1 - D^{-n}(u)$ is convex. Now it is easy to show that $\varphi(u)$ is also convex, using the uniform convergence of $\varphi_n(u)$ on any compact $[a,b]$ in $]0,1[$.

- For all n, φ_n is strictly decreasing with u: namely D is increasing, therefore D^{-1} and also D^{-n}. Therefore $\varphi(u)$ is strictly decreasing in u being a limit of strictly decreasing functions.

Thus φ possesses all the properties required for a generator of an Archimedean copula.

Given D, we define φ as in Eq. (4.34) and the Archimedean copula is then:

$$C(u,v) =) = \lim_{n\to\infty} D^n(D^{-n}(u) + D^{-n}(v) - 1).$$

Remark : if the condition (5) is not fulfilled , that is if

$$\forall u \in (0,1), \quad \lim_{n\to\infty} 2^n(D^{-n}(u) - 1) = \infty$$

one can still construct a generator φ for an Archimedean copula : Sungur and Yang [212] construct it pointwise.

4.6.5.1 *Fitting an observed distribution with an Archimedean copula*

Using the results of the preceding Section, and the property that a diagonal copula is the cumulative distribution function of $W = max(U,V)$ (Section 4.1), we can fit an observed distribution to an Archimedean copula. The approach described below is due to Sungur and Yang [212].

-Given a sample (X_i, Y_i), $i = 1, ..., n$, we can construct the sample $W_i = max(F_{1n}(X_i), F_{2n}(Y_i))$, and estimate $G_n(w)$, the empirical distribution function of W_i.

We ought also to assume that $G_n(w)$ is always less than w. One can, in a given Archimedean family of copulas indexed by a parameter θ, estimate this parameter, applying, for example, the method of moments. Explicitly we shall choose the family 1 (Section 4.6.2) with the generator

$$\varphi(u) = (1-u)^\theta, \ \theta > 1 .$$

In this case the derivative of the diagonal copula $D(w) = \varphi^{-1}(2\varphi(w))$ is a constant. Namely

$$D'(w) = \frac{2\varphi'(w)}{\varphi'(\varphi^{-1}2\varphi(w))} = \frac{2\theta(1-w)^{\theta-1}}{\theta.2^{\frac{\theta-1}{\theta}}(1-w)^{\theta-1}} = 2^{1/\theta} .$$

Therefore $D(w)$ has here a uniform distribution on a part of $(0,1)$. The expectation of W is:

$$E(W) = \int_{[1-2^{-\frac{1}{\theta}},1]} 2^{\frac{1}{\theta}} w dw = 1 - 2^{-\frac{1}{\theta}-1}.$$

This expectation is estimated by $\bar{W} = \frac{\sum W_i}{n}$, and one deduces the estimator of θ to be

$$\hat{\theta} = -\frac{log2}{log2(1-\bar{W})}.$$

The difficulty with this method is that, first by we have no criteria to choose *a priori* one family or another (although we could compare *a posteriori* the empirical distribution $D_n(w)$ with the fitted distribution $\tilde{D}_\theta(u)$). Second the expression of $D'(w)$ as a function of the parameter may not be so simple as in the example and consequently the expectation of W can be intractable. The method described in the next Section, due to Genest and Rivest [83], allows us to fit the observed distribution with any Archimedean copula in a simple manner. The comparison between the empirical and the fitted distributions is carried out by means of a χ^2 test. In the last chapter we shall provide criteria for choosing the family.

4.6.6 *Characterization of an Archimedean copula by the cumulative distribution function of $Z = C(U,V)$*

The key results (due to Genest and Rivest[83]) in this section are:

(1) The function $K(z) = z - \frac{\varphi(z)}{\varphi'(z)}$ is the cumulative distribution function of the variable $Z = C(U,V)$. Therefore knowing $K(z)$, one can

in principle retrieve the function $\varphi(z)$ and hence the Archimedean copula.

(2) The function $K(z)$ can be estimated by means of the empirical distribution function $K_n(z_i)$, where Z_i is the proportion of the pairs (X_j, Y_j), which are less than or equal to the pair (X_i, Y_i).

(3) The empirical function $K_n(z)$ can be fitted by the distribution functions $K_{\hat{\theta}}(z)$ of any family of Archimedean copulas, where the parameter θ is estimated in such a manner that the fitted distribution has a coefficient of concordance (τ) equal to the corresponding empirical coefficient (τ_n).

More precisely, Genest and Rivest [83] prove the following proposition:

Proposition: Let (U, V) be a random vector from an Archimedean copula, with generator φ . Set $W = \frac{\varphi(U)}{\varphi(U)+\varphi(V)}$ and $Z = C(U, V)$. Then

(1) W is distributed uniformly on (0,1);
(2) Z is distributed as $K(z) = z - \lambda(z)$, where $\lambda = \frac{\varphi(z)}{\varphi'(z)}$;
(3) Z and W are independent.

These results can be used in other contexts as well, for example for generating a sample (U, V) from a given Archimedean distribution.

Proof of the proposition: Assume that C is absolutely continuous, and let $g(w, z)$ be the density of (U, V) and $G(z, w) = P(Z \le z, W \le w)$, then

$$G(z, w) = \int_0^z \int_0^w g(w, z) dw dz = \int\int c(u, v) |\frac{\partial(u, v)}{\partial(z, w)}| du dv ,$$

where $c(u, v) = \frac{\partial^2 z}{\partial u \partial v}$ is the density of the copula and $\frac{\partial(u,v)}{\partial(z,w)}$ is the Jacobian of the transformation $(u, v) \to (z, w)$. Here we have $\varphi(u) = w.\varphi(z)$, $\varphi(v) = (1 - w)\varphi(z)$ and the Jacobian equal to $\frac{\varphi(z)\varphi'(z)}{\varphi'(u)\varphi'(v)}$. Proceeding as in Section 4.6.1 (Eq. 4.31), we have :

$$\frac{\partial^2 z}{\partial u \partial v} = \frac{-\varphi'(u)\varphi'(v)\varphi''(z)}{\varphi'(z)^3}.$$

Hence:

$$\begin{aligned} G(z, w) &= \int_0^z \int_0^w \frac{-\varphi'(u)\varphi'(v)\varphi''(z)}{\varphi'(z)^3}.(-\frac{\varphi(z)\varphi'(z)}{\varphi'(u)\varphi'(v)}) = w\left[z - \frac{\varphi(z)}{\varphi'(z)}\right]_0^z \\ &= wK(z). \end{aligned}$$

The last relationship proves the proposition.

4.6.7 *Archimedean copulas with two parameters*

It is easy to prove that by means of a generator φ of an Archimedean copula, it is possible to construct other families of generators [161]. Indeed if Φ is the family of all the Archimedian copulas then:

(1) If $\varphi \in \Phi$, then φ_β is such that $\varphi_\beta(t) = (\varphi(t))^\beta \in \Phi, \ \forall \beta \geq 1$.
(2) If $\varphi \in \Phi$, then φ_α is such that $\varphi_\alpha(t) = (\varphi(t^\alpha)) \in \Phi, \ \forall 0 < \alpha \leq 1$.
(3) Using these two properties successively, we obtain that $\varphi_{\alpha\beta}(t) = \varphi_\beta \circ \varphi_\alpha(t)$ is also the generator of an Archimedean copula.

Examples:

(1) For $\varphi(t) = t^{-1} - 1$ we obtain $\varphi_\alpha(t) = t^{-\alpha} - 1$, for $\alpha > 0$ generates an Archimedian family (Clayton family).
(2) For $\varphi(t) = -ln(t)$ and $\beta \geq 1$, we obtain $\varphi_\beta(t) = (-ln(t))^\beta$, which corresponds to Gumbel-Hougaard family.
(3) The generator $\varphi_{\alpha\beta}(t) = (t^\alpha - 1)^\beta$ with $\alpha \geq 0$ and $\beta \geq 1$, allows us to construct the Archimedean copula:

$$C_{\alpha,\beta}(u,v) = \{[(u^{-\alpha} - 1)^\beta + (v^{-\alpha} - 1)^\beta]^{\frac{1}{\beta}} + 1\}^{-\frac{1}{\alpha}} .$$

4.7 Archimax Copulas

Recently Capéraà *et al.* [38] have defined a new family of copulas for which Archimedean copulas and extreme value copulas are particular cases.

4.7.1 *Extreme value distribution and extreme value copula*

Let $(U_1, V_1),\ldots ,(U_n, V_n)$ be a random sample from an arbitrary copula C and let $U_{max} = max(U_1, \ldots, U_n)$ and $V_{max} = max(V_1, \ldots, V_n)$ be the two maxima in the sample. The extreme value distribution is the limiting distribution, if it exists, when n tends to infinity of (U_{max}, V_{max}). However:

$$P(U_{max} \leq u, V_{max} \leq v) = C^n(u,v)$$

and the marginals of this distribution are:

$$P(U_{max} \leq u) = u^n .$$

Hence the copula associated with the extreme value distribution of C is:

$$C_{max}(u,v) = lim_{n\to\infty} C^n(u^{\frac{1}{n}}, v^{\frac{1}{n}}) .$$

In that case we say that C belongs to the domain of attraction of C_{max}.

4.7.2 *Definition of Archimax copulas*

Following the work of Pickands [168], Capéraà *et al.* [37] use the general form of a bivariate extreme value copula :

$$C_A(u,v) \equiv \exp\left[\log(uv).A\left(\frac{\log(u)}{\log(uv)}\right)\right] , \tag{4.35}$$

where A is a convex function $[0,1] \mapsto [1/2,1]$ such that $max(t, 1-t) \leq A(t) \leq 1$ for all $t \in [0,1]$.

The family of Archimax copula is then defined as:

$$C_{\varphi,A}(u,v) = \varphi^{-1}\left[\{\varphi(u)+\varphi(v)\}A\left(\frac{\varphi(u)}{\varphi(u)+\varphi(v)}\right)\right],$$

where A is the convex function defined above and φ is the generator of an Archimedean copula.

If $A \equiv 1$, we retrieve the Archimedean copulas, and if $\varphi(t) = -ln(t)$, we retrieve the extreme value copula with the Capéraà formulation (Eq. 4.35).

Remark: This procedure to generate a bivariate copula is a particular case of Marshall and Olkin's generalization (Section 4.5.6, Eq. 4.22), where the function K is here the extreme value copula C_{A^*}, and the mixture distribution G has for its Laplace transform the generator φ^{-1}.

4.7.3 *Construction of bivariate distributions belonging to a predetermined domain of attraction*

The aim of Capéraà *et al.* is then to construct families of Archimax copulas belonging to the domain of attraction of a predetermined extreme value copula C_{A^*}. For this an additional condition is needed on the generator φ : $\varphi(1-\frac{1}{t})$ must be *regularly varying at* ∞ with degree $-m$ for some $m \geq 1$. That is :

$$lim_{t\to\infty}\frac{\varphi(1-\frac{s}{t})}{\varphi(1-\frac{1}{t})} = s^{-m},\ \forall\, s > 0.$$

One then writes $\varphi(1-\frac{1}{t}) \in RV_{-m}$.

Proposition (Capéraà *et al.*): The Archimax copula $C_{\varphi,A}$, with the function $A \in RV_{-m}$ belongs to the domain of attraction of C_{A^*}, where :

$$A^*(t) = (t^m + (1-t)^m)^{1/m} A^{1/m}\left(\frac{t^m}{t^m + (1-t)^m}\right).$$

This attractor may be regarded as an Archimax distribution with generators A^* and $\varphi(t) = -ln(t)$ (or with generators A and $\varphi^*(t) = (-ln(t))^m$). Furthermore, A and A^* coincide iff $m = 1$.

Conversely, given $\varphi(1 - 1/t) \in RV_{-m}$ and A^*, one can find a function A by means of the relation:

$$A(t) = (t^{1/m} + (1-t)^{1/m})^m . \left(A^*(\frac{t^{1/m}}{t^{1/m} + (1-t)^{1/m}})\right)^m .$$

However condition on m is imposed to assure that $C_{\varphi,A}$ belongs to the domain of attraction of C_{A^*}.

4.7.4 *Examples*

(1) The Gumbel-Hougaard family
It is the only family which is both Archimedean and extreme value copula.

(2) $A_\theta(t) = \theta.t^2 - \theta.t + 1$, $0 \leq \theta \leq 1$ and $\varphi(t) = t^{1-\alpha} - 1$ $\alpha > 1$.
Here $A_\theta(t)$ is known as the generator of the Tawn mixed model and $\varphi(t)$ is the Clayton's generator.

(3) $A_\theta(t) = \theta t^2 - \theta.t + 1$, $0 \leq \theta \leq 1$ and $\varphi(t) = (1 - t^\alpha)^{1/\alpha}$, $0 < \alpha < 1$ (which is a generator constructed from the example 1 of Section 4.6.2 $\varphi(t) = (1-t)^\alpha$ and transformed using method developed in section 4.6.7).

4.8 Copulas with Discontinuity Constraints

Wei *et al* [223] have studied copulas whose probability measure is concentrated on a set of measure zero, called "piecewise additive copulas" and copulas whose density functions are piecewise constant (that are locally independent) which they call "piecewise quadratic copulas" and a particular case of the preceding copulas for which the density function is zero on some "holes" of the square unit and piecewise constant elsewhere.

4.8.1 *Piecewise additive copulas*

Suppose that the unit square is partitioned into measurable closed sets: $[0,1]^2 = \bigcup_{i=1}^{\infty} A_i$ whose boundaries DA_i are piecewise differentiable curves with finite length and $A_i \bigcap A_j \subset DA_i \bigcap DA_j$ for all $i \neq j$.

Theorem: The joint distribution of (U, V) is singular iff the copula is piecewise additive, i.e. on each partition set A_i, we have

$$C(u,v)|_{A_i} = C_1(u) + C_2(v),$$

where $C_1(u)$ and $C_2(v)$ are some increasing functions.

Proof

(1) Necessity
If the joint distribution is singular, then there exists a partition on the unit square such that, the probability measure is concentrated along the boundaries of the partition set. Thus within each partition set the increment of the cumulative distribution function is the sum of increments of conditional marginal distributions :

$$\begin{aligned} C(u+\Delta u, v+\Delta v)|_{A_i} &= [C(u+\Delta u, v) - C(u,v)] \\ &+ [C(u, v+\Delta v) - C(u,v)] \\ &= C_1(\Delta u) + C_2(\Delta v). \end{aligned}$$

(2) Sufficiency
Since an additive copula possesses no density and since the cumulative distribution function increases by increments, the probability measure is concentrated on the boundaries of the partition set.

Example: Let (X, Y) be uniformly distributed in L_1 norm on the unit sphere $\{(x,y) : |x| + |y| = 1\}$. The corresponding copula is then concentrated along the lines $y = x \pm \frac{1}{2}$ and $y = -x + 1 \pm \frac{1}{2}$. Therefore there exists a partition of the square unit into five regions :

$$C(u,v) = \begin{cases} \frac{u+v}{2} - \frac{1}{4} & \text{if } |u-\frac{1}{2}| + |v-\frac{1}{2}| \leq \frac{1}{2} \\ & \text{if } |u-\frac{1}{2}| + |v-\frac{1}{2}| > \frac{1}{2} \\ 0 & 0 \leq u < \frac{1}{2} \text{ and } 0 \leq v < \frac{1}{2} \\ u & 0 \leq u < \frac{1}{2} \text{ and } \frac{1}{2} \leq v < \frac{1}{2} \\ v & \frac{1}{2} \leq u < 1 \text{ and } 0 \leq v < \frac{1}{2} \\ u+v-1 & \frac{1}{2} \leq u < 1 \text{ and } \frac{1}{2} \leq v < 1. \end{cases}$$

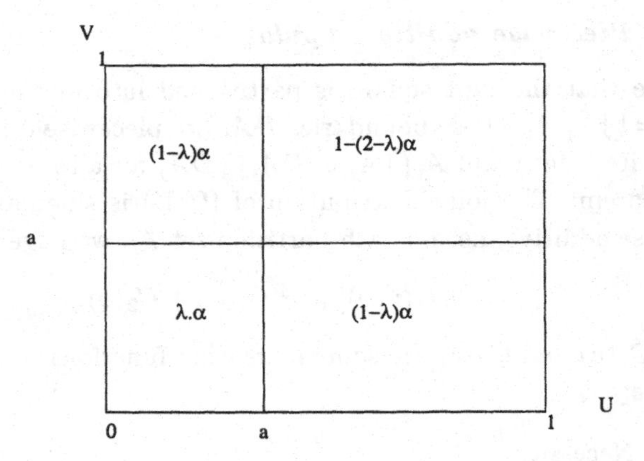

Fig. 4.1 A piecewise quadratic copula.

Thus $C(u, v)$ is piecewise linear.

Remark : The uniform distribution on the circle $\{x^2 + y^2 = 1\}$ has the same copula ([223]).

4.8.2 *Piecewise quadratic copulas*

A piecewise quadratic copula has a piecewise constant density. For example: let $0 < \lambda < 1$ and $0 \leq \alpha \leq \frac{1}{2-l}$, and a partition of the unit square into four rectangular regions as given in Figure 4.1.

4.8.3 *Quadratic copulas with holes*

Definition: Let (U, V) be a continuous random vector over the unit square $J = [0, 1]^2$. Let S be the support of the density function. Let us call DM, the boundary of a set M. A connected set A is a hole if $A \subset \mathcal{J} - \mathcal{S}$ and $DA \subset (DJ \bigcup DS)$.

A simple example: consider the family of copulas with a single hole A and the density constant over $\mathcal{U} - \mathcal{A}$. If the hole has width α and height β then the density over $\mathcal{J} - \mathcal{A}$ is given by $\frac{1}{1-\alpha\beta}$ (Fig. (4.2)).

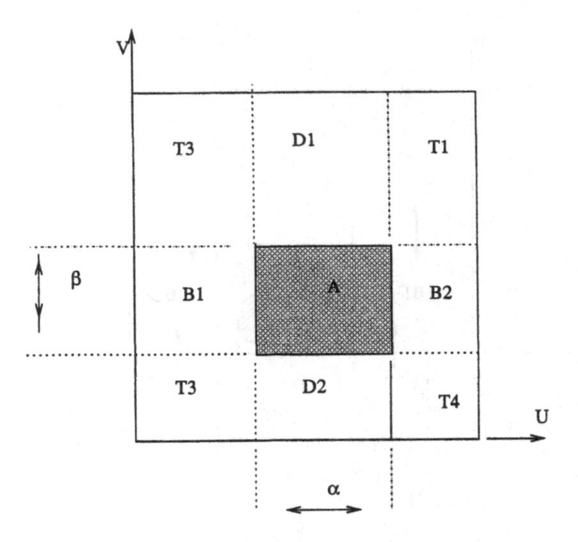

Fig. 4.2 A rectangular hole within a copula.

4.8.3.1 *Admissible rectangles*

Wei *et al.* propose a heuristic method (inspired by a conterexample of Stoyanov [211]) for constructing a copula with holes. Since the marginals of a copula are uniforms, the idea is to shift the omitted mass of holes along one axis, next along the other axis and again along the first one as schematically indicated in Fig. (4.3).

Before describing the method we have to define the concept of an admissible rectangle, i.e. a rectangle possessing the holes. First we shall consider a single rectangular hole of with width α and height β.

From the definition of a hole, the probability measures of the rectangles D_1, D_2, B_1, and B_2 must be positive. We shall denote them by d_1, d_2, b_1, b_2. Since the marginals are uniform, we have:

$$1 \times \alpha = d_1 + d_2$$

$$1 \times \beta = b_1 + b_2.$$

However the probability measures of the rectangles T_2, T_1, T_3 and T_4 must

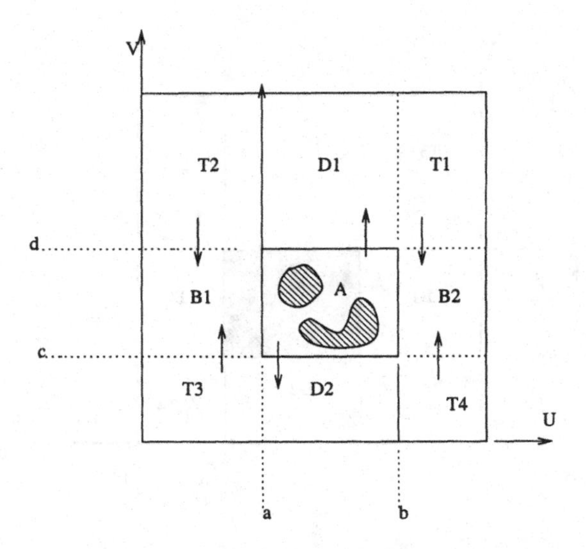

Fig. 4.3 Holes inside an admissible rectangle

also be strictly positive, so that:

$$1 - \alpha - \beta > 0$$

From these constraints, we deduce that:

$$\alpha.\beta < max(\alpha(1-\alpha), \beta(1-\beta)) < \frac{1}{4}.$$

A rectangle which satisfies these constraints is called admissible.

If α and β satisfy these constraints, we can construct a copula with a probability mass equal to α on the two rectangles (D_1, D_2), a mass equal to β on the two rectangles (B_1, B_2) and the remainder of the mass $1 - \alpha - \beta$ situated on T_1, T_2, T_3 and T_4.

4.8.3.2 *The squeeze algorithm*

We shall consider now more generally a construction of copulas with holes that are contained within an admissible rectangle.

The construction algorithm is as follows (see Fig. 4.3):

(1) Start from the uniform distribution on the unit square. Suppose that the mass measure of holes is s. Set the density to zero over the holes.
(2) Distribute this mass over $D_1 \bigcup D_2$ such that the marginal density along u will remain 1 for $u \in [0,1]$.
(3) Remove the mass s from the T_i , $i = 1, ..., 4$ so the marginal density along v will remain 1, for $v \in [0,c] \bigcup [c,1]$.
(4) Distribute the mass s over $B_1 \bigcup B_2$, so the marginal density along u for $u \in [0,a] \times [b,1]$ is 1 and along v for $v \in [c,d]$ is also 1.

4.9 Copulas with More than Two Variables

It is quite tempting to use the machinery of copulas to construct multivariate distributions with given marginals for m ($m > 2$) variables. There are many ways to generalize with respect to the marginals that we have decide to fix: fixing m univariate marginals and searching for a mapping $[0,1]^m \mapsto [0,1]$, which defines a dependence structure, or fixing two marginals $F_p(x_1, x_2, ..., x_p)$ and $G_q(y_1, y_2, ..., y_q)$ with $p+q = m$ and searching for a mapping $[0,1]^2 \mapsto [0,1]$, or fixing some 2–dimensional marginals, etc. However there are compatibility constraints on the marginals. For example in the class of 3-dimensional copulas, the marginals C_{12}, C_{13} and C_{23} overlap and are therefore not completely independent from one another. Furthermore a given family of m–copulas ($m > 2$) does not attain its lower Fréchet bounds.

The tool of copulas is less universal in the case of m ($m \geq 3$) variables than in the case of two. Specifically Genest *et al.* [84] have shown that given C a 2-copula and two marginals $F_m(x_1, x_2, ..., x_m)$, and $G_n(y_1, y_2, ..y_n)$, the only copula which satisfies that $H(x_1, ..., x_m, y_1, ..., y_n) = C(F_m(x), G_n(y))$ is a cumulative distribution function with the marginals F and G for *any* F and G, is the independent copula. To show this result Genest *et al.* utilize an example where F, of dimension 2, is precisely the Fréchet lower bound $max\{0, X_1 + x_2 - 1\}$. However if we impose some restrictions on the marginals we may hopefully be able to construct families of multivariate distributions.

4.9.1 *m-dimensional Archimedean copulas*

Jouini and Clemen [120] propose to iterate the bivariate Archimedean copulas to obtain any m–dimensional Archimedean copula using the property of associativity (Ling's theorem, see Section 4.6.3).

The procedure is as follows. Let C be a given two-dimensional copula. Define

$$C_2 = C$$

and recursively

$$C_m = C(C_{m-1}, u_m),\ m > 2$$

This procedure works well in the case of frailty models, as we have already seen in Section 4.5.6, (Eq. 4.24). If φ^{-1} is a Laplace transform, the expression C_m becomes:

$$C_m(u_1, ..., u_m) = \varphi^{-1}\left(\varphi(u_1) + .. + \varphi(u_m)\right). \tag{4.36}$$

Moreover, as it was shown by Schweizer and Sklar [196], in order to extend to all n–dimensional distributions on the positive region, φ^{-1} must be proportional to a Laplace transform. When it is not the case, i.e. for Archimedean copulas which are not frailty models, representation (4.36) does not apply.

For example, in the case of Clayton's family (Section 4.6.2):

$$C_n(u_1, ..., u_m) = [u_1^{1-\alpha} + ... + u_m^{1-\alpha} - m + 1]^{\frac{1}{1-\alpha}},\ \alpha > 1.$$

Now for $m = 2$, the lower Fréchet bound $C(u, v) = max(0, u + v - 1)$ is Archimedean (with the generator $\varphi(t) = 1 - t$), however iterating we obtain

$$C(u_1, u_2, u_3) = max(u_1 + u_2 + u_3 - 2, 0)$$

which is not a cumulative distribution function (see, e.g., Problem 748 (1976) American Mathematical Monthly, 83 and Solution : American Mathematical Monthly, 85, (1978) p. 393).

4.9.1.1 *An application*

Meester and MacKay [154] have proposed to use Frank copulas model for fitting models to clustered binary data. The size of the cluster is n (in general $n > 2$). This is possible since Frank model is obtained by Laplace transform (Eq. (4.36)). The association parameter α of the Frank copula model provides a measure of the strength of the within-cluster association. The advantage of Frank model is that even with binary marginals, the parameter α is almost independent of the marginal probabilities, which are also modeled. Tests on the parameters are performed using maximum likelihood method and are computationally straightforward.

Trégoüet *et al.* have applied this idea to modelling the joint distribution of a binary trait (a disease status) within families. They decompose a family into two sets (parents and offspring), each of them characterized by an association parameter (α_{FM} and α_{SS} respectively). The marginals probabilities are modeled by means of a logistic representation. They perform a segregation-linkage analysis of levels of plasma angiotensin converting enzyme (ACE) dichotomized into two classes. The key steps are the following:

(1) Consider a family in which a binary trait (disease status) is measured. Let $\mathbf{Y} = (Y_F, Y_M, Y_1, ..., Y_n)$ the vector of the trait for the father (F), the mother (M), and the n children. Similarly $\mathbf{x} = (x_F, x_M, x_1 ..., x_n)$ be a vector of covariates of the family.

(2) Suppose that, conditionally on an individual's own covariates, an individual's status is independent of the covariates of the other family members. The joint probability of the trait given the covariates can then be decomposed into two probabilities :

$$P(\mathbf{Y}/\mathbf{x}) = P(Y_F, Y_M/x_F, x_M).P(Y_1, ..., Y_n/Y_F, Y_M, x_1..x_n)$$

(3) The two probabilities are modeled by two different Frank's models (with parameters α_{FM} and α_{SS} respectively) since (Y_F, Y_M) and $(Y_1, ..., Y_n)$ are considered each to be equicorrelated data. For example for the children we have:

$$\begin{aligned} &P(Y_1 \le y_1, ..., Y_n \le y_n) = \\ &C_{\alpha_{SS}}(F_1(y_1), ..., F_n(y_n)) = \\ &log_\alpha\left(1 + (\alpha - 1)\Pi_{i=1}^n\left(\frac{\alpha^{F_i(y_i)} - 1}{\alpha - 1}\right)\right). \end{aligned}$$

(4) The marginal distribution functions are modeled by a logistic representation.

For example, for a mother we have

$$F_F(Y_F/x_F) = \begin{cases} 0 & \text{if } Y_F = -1 \\ \frac{1}{1+e^{\lambda+\beta.x_F}} & \text{if } Y_F = 0 \\ 1 & \text{if } Y_F = 1 \end{cases}$$

where λ is the baseline hazard and β is the vector of marginal regression parameters of covariates.
For a sibling:

$$F_F(Y_i/x_i, y_F, y_M) = \begin{cases} 0 & \text{if } Y_i = -1 \\ \frac{1}{1+e^{\lambda+\beta.x_i+\gamma_{FO}\cdot y_F+\gamma_{MO}\cdot y_M}} & \text{if } Y_i = 0 \\ 1 & \text{if } Y_i = 1 \end{cases}$$

where γ_{FO} and γ_{MO} are the regression coefficients, for the parents phenotypes, that characterize the familial aggregation between the parents and the offspring.

(5) The joint probability mass function, for example for the pair (Y_F, Y_M) is then formally written as:

$$\begin{aligned} P(Y_F, Y_M/x_F, x_M) = \quad & C_{\alpha_{FM}}\left(F_F(Y_F/x_F), F_M(Y_M/x_M)\right) \\ & -C_{\alpha_{FM}}\left(F_F(Y_F/x_F), F_M(Y_M - 1/x_M)\right) \\ & -C_{\alpha_{FM}}\left(F_F(Y_F - 1/x_F), F_M(Y_M/x_M)\right) \\ & +C_{\alpha_{FM}}\left(F_F(Y_F - 1/x_F), F_M(Y_M - 1/x_M)\right) \end{aligned} \tag{4.37}$$

Formulas for the children are similar.

The association parameter α_{SS} in the model for the sibs characterizes the residual aggregation between sibs after controlling for the parents-offspring effects (which is represented by the parameters γ_{FO} and γ_{MO}). It can be due, for example, to the shared environmental factors.

Actually, the model is more complicated and incorporates a major-gene effects. Therefore the two association parameters α_{FM} and α_{SS} correspond to a residual effect between the spouses, and between the sibs after controlling the major gene effects.

Conclusions of this analysis are similar to those reported in an earlier analysis where the data of the ACE levels were not dichotomized, however there is a loss of power due to the dichotomization and the residual effects of association could not be shown to be significant.

4.9.2 *Generation of a 3-dimensional copula from its 2-dimensional marginals*

4.9.2.1 *Compatibility of marginals*

Suppose that we are searching for a copula C_{123}, with fixed marginals C_{12}, C_{13}, C_{23}. We have to verify (using a shorthand notation) that :

$$C_{12} = \int C_{1/3} C_{2/3} du_3$$

where $C_{1/3}$ and $C_{2/3}$ are the conditional distributions given u_3 and where the integration is taken on the domain of definition of u_3. The two other relations obtained by permuting the subscripts $\{1, 2, 3\}$ ought to be verified as well.

4.9.2.2 *Truncation invariance*

Chakak and Koehler [40], and subsequently Sungur [213] defined a class of three-dimensional truncation invariant copulas which are determined solely by their two-dimensional marginals. Using this procedure, it is therefore possible to construct a single 3–dimensional copula, choosing 2–dimensional marginals provided that the marginals are compatible.

A 3-dimensional copula is truncation invariant in one variable, if the dependence structure between the two other variables is not affected by the truncation of this variable: for all $a_3 \in [0, 1]$ the distribution function of $(U_1, U_2)/U_3 \leq a_3$ is the same as the marginal distribution function of (U_1, U_2). In that case the 3-dimensional copula satisfies:

$$C_{123}(u_1, u_2, u_3) = C_{12}\left[\frac{C_{13}(u_1, u_3)}{u_3}, \frac{C_{23}(u_2, u_3)}{u_3}\right] u_3 \,. \tag{4.38}$$

Namely, the joint distribution of $(U_1^{tr}, U_2^{tr}) = [(U_1, U_2)/U_3 \leq a_3]$ is:

$$F(U_1^{tr}, U_2^{tr}) = \frac{P(U_1 \leq u_1, U_2 \leq u_2, U_3 \leq a_3)}{P(U_3 \leq a_3)} \,. \tag{4.39}$$

The marginal distributions of $U_i^{tr} = U_i/U_3 \leq a_3$, $i = 1, 2$, are:

$$F_i^{tr}(u_i) = \frac{P(U_i \leq u_i, U_3 \leq a_3)}{P(U_3 \leq a_3)} = \frac{C_{i3}(u_i, a_3)}{a_3} \,.$$

If C_{a_3} denotes the copula associated with (U_1^{tr}, U_2^{tr}) namely:

$$F(U_1^{tr}, U_2^{tr}) = C_{a_3}(F_1(U_1^{tr}, F_2(u_2^{tr}))) = C_{a_3}\left(\frac{C_{13}(u_1, a_3)}{a_3}, \frac{C_{23}(u_2, a_3)}{a_3}\right),$$

then :

$$\begin{aligned} C_{123}(u_1, u_2, a_3) &= P(U_1 \leq u_1, U_2 \leq u_2, U_3 \leq a_3) \\ &= F(U_1^{tr}, U_2^{tr})a_3 \\ &= C_{a_3}\left(\frac{C_{13}(u_1, a_3)}{a_3}, \frac{C_{23}(u_2, a_3)}{a_3}\right). \end{aligned}$$

If this result holds for all a_3, then Equation (4.38) is valid.

If this property is also valid when conditioning on the two other variables, the copula will be truncation invariant.

Examples:

(1) pairwise independent bivariate copula :

$$C_{ij} = u_i.u_j \;\; i \neq j \;\in 1, 2, 3.$$

In this case

$$C_{123}(u_1, u_2, u_3) = C_{jk}(\frac{C_{ij}(u_i, u_j)}{u_i}, \frac{C_{ik}(u_i, u_k)}{u_i}).u_i = u_1 u_2 u_3.$$

Hence for this construction pairwise independence implies mutual independence.

(2) Plackett's family (Chakak and Koehler [40]):
In the Plackett's copula [169], the odds-ratio :

$$\alpha_{12} = \frac{C_{12}(u, v)\bar{C}(u, v)}{(u - C_{12}(u, v))(v - C_{12}(u, v))}$$

is constant. The associated copula (cf Section 4.4 Eq. (4.13)) is :

$$C_{12}(u, v) = \frac{G(u, v) - \sqrt{G^2(u, v) - 4\alpha_{12}(\alpha_{12} - 1)uv}}{2(\alpha_{12} - 1)} \quad \alpha_{12} > 0$$

with $G(u, v) = 1 + (\alpha_{12} - 1)(u + v)$. Construction of the trivariate copula by conditioning on u_3 gives :

$$C_{123}(u, v) = C_{12}\left[\frac{C_{13}(u_1, u_3)}{u_3}, \frac{C_{23}(u_2, u_3)}{u_3}\right] u_3$$

$$= \frac{G_{12.3} - \sqrt{G_{12.3}^2 - 4\alpha_{12}(\alpha_{12}-1)C_{13}C_{23}}}{2(\alpha_{12}-1)} \tag{4.40}$$

where $G_{12.3} = u_3 + (\alpha_{12}-1)(C_{13}+C_{23})$. One can also obtain this copula by stipulating that the conditional odds-ratio given u_3 is constant for all u_1, u_2 .This condition yields

$$\alpha_{12/3} = \frac{C_{123}(u_3 - C_{13} - C_{23} + C_{123})}{(C_{13} - C_{123})(C_{23} - C_{123})}.$$

Solving this equation, by stipulating that $\alpha_{12/3}$ is constant, we then retrieve Eq. (4.40). However, for the Plackett's copula, conditioning on other variables than u_3 does not result in the same copula.

(3) Clayton's family :
This family satisfies truncation invariance. Namely

$$C(u_1, u_2, u_3) = (u_1^{1-\alpha} + u_2^{1-\alpha} + u_3^{1-\alpha} - 2)^{\frac{1}{1-\alpha}} \tag{4.41}$$

can be rewritten as:

$$C(u_1, u_2, u_3) =$$

$$\left(\frac{1}{[\frac{(u_1^{1-\alpha}+u_3^{1-\alpha}-1)^{1/1-\alpha}}{u_3}]^{\alpha-1}} + \frac{1}{[\frac{u_2^{1-\alpha}+u_3^{1-\alpha}-1)^{1/1-\alpha}}{u_3}]^{\alpha-1}} - 1 \right)^{\frac{1}{1-\alpha}} u_3 ,$$

which is

$$C_{12}(\frac{C_{13}(u_1, u_3)}{u_3}, \frac{C_{23}(u_2, u_3)}{u_3}).u_3 .$$

This equality remains valid if we permute the indices $\{1, 2, 3\}$.

(4) Three parameters Clayton's family:
It is possible to generalize the preceding formula with α_i, $i = 1, 2, 3$, being associated with the bivariate copulas C_{12},C_{13} and C_{23}. If

$$C_{ij}(u_i, u_j) = \left(u_i^{1-\alpha_i} + u_j^{1-\alpha_i - 1}\right)^{\frac{-1}{1-\alpha_i}} , \ \alpha_i > 1 .$$

Provided that the marginals C_{ij}, $i \neq j$ are compatible, one can construct the copula :

$$C(u_1, u_2, u_3) =$$

$$\left[(u_1^{1-\alpha_1} + u_2^{1-\alpha_1} - 1)^{\frac{1-\alpha_3}{1-\alpha_1}} + (u_1^{1-\alpha_2} + u_3^{1-\alpha_2} - 1)^{\frac{1-\alpha_2}{1-\alpha_3}} - u_1^{1-\alpha_3}\right] \quad (4$$

Note that Eq. (4.42) was obtained by conditioning on u_1. Analogous expressions could be obtained by conditioning on u_2 and u_3.

(5) Lower Fréchet bound copula :
For $C^-(u_i, u_j) = \max(u_i + u_j - 1, 0)$, we can construct

$$\begin{aligned} C(u_1, u_2, u_3) &= C^-\left(\frac{C^-(u_1, u_3)}{u_3}, \frac{C^-(u_2, u_3)}{u_3}\right) \\ &= \max\left[\max(u_1 + u_3 - 1, 0) + \max(u_2 + u_3 - 1, 0) - u_3, 0\right]. \quad (4.4 \end{aligned}$$

This procedure can of course be generalized to obtain m-dimensional copulas ($m > 3$) using only the two-dimensional marginals, and the property of truncation invariance. We note that since the formal three-dimensional lower Fréchet bound is not attained Eq. (4.43) may be viewed as an alternative version.

4.9.3 *Linkages*

In the case of two variables, a copula allows us to separate the effects of the univariate marginals from the effect of dependence. Using the same idea and being inspired by the setwise dependence, Li *et al.* [143] have proposed extending this approach to the case when $(\mathbf{X}_1, \mathbf{X}_2, ..., \mathbf{X}_k)$ are k random vectors of dimensions m_1,m_2,...,m_k respectively with the marginals F_1,F_2,...,F_k. For each random vector $\mathbf{X}_i = (X_{i1}, X_{i2}, ..., X_{ik_i})$, they associate a vector $\mathbf{U}_i = (U_{i1}, U_{i2}, ..., U_{i_{k_i}})$ where the U_{i_j} are independent and uniform on $[0, 1]$. The linkage is then defined as the joint distribution function L of the vector $(U_1, ..., U_k)$. Thus the linkage is useful in studying the dependence properties *between* the $\mathbf{X}_i$ separate from the dependence properties *within* each $\mathbf{X}_i$. However decomposition of each vector $\mathbf{X}_i$ into $\mathbf{U}_i$ is not unique, it depends on the order of the subscript $i_1, ..., i_{k_i}$, and therefore the linkage L is also not unique.

How does this construction work? Let $\mathbf{X} = (X_1, ..., X_k)$ be a k-dimensional vector with an absolutely continuous joint distribution function F and the marginals F_i , $i = 1, ..., k$. The procedure is as follows :

(1) $U_1 = F_1(X_1)$

(2) $U_2 = F_{2/1}(X_2/X_1 = x_1)$
where $F_{2/1}$ represents the conditional distribution function of X_2 given $X_1 = x_1$.
(3) $U_i = F_{i/1,2,i-1}(X_i/x_1, ..., x_{i-1})$

In this construction the variables U_i are independent and uniformly distributed.

Example : Let $\mathbf{X} = ((W_1, W_2), (Z_1, Z_2))$ be a four-dimensional normal random vector with mean 0 and the correlation matrix given by :

$$\Sigma = \begin{pmatrix} 1 & \rho_w & \rho & \rho \\ \rho_w & 1 & \rho & \rho \\ \rho & \rho & 1 & \rho_z \\ \rho & \rho & \rho_z & 1 \end{pmatrix}.$$

The constraint on the parameters is :
$-1 \leq \rho \leq \frac{1}{2}\sqrt{(1+\rho_w)(1+\rho_z)}$ in order to assure that Σ be a semi-definite positive matrix. We have : $W_1 \sim N(0,1)$; $W_2/(W_1 = w_1) \sim N(\rho_w.w_1, 1-\rho_w^2)$, $Z_1 \sim N(0,1)$, $Z_2/(Z_1 = z_1) \sim N(\rho_z.z_1, 1-\rho_z^2)$. Thus the joint distribution function L is :

$$\begin{pmatrix} U_1 \\ U_2 \\ U_3 \\ U_4 \end{pmatrix} = \begin{pmatrix} \Phi(W_1) \\ \Phi(\frac{W_2-\rho_w.W_1}{\sqrt{1-\rho_w^2}}) \\ \Phi(Z_1) \\ \Phi(\frac{Z_2-\rho_z.Z_1}{\sqrt{1-\rho_z^2}}) \end{pmatrix}$$

where $\Phi(.)$ is the standard normal distribution function. To arrive at a more indicative form of L, Li *et al.* [143] present the distribution function of the vector $(\Phi^{-1}(U_1), \Phi^{-1}(U_2), \Phi^{-1}(U_3), \Phi^{-1}(U_4))$ which is a normal distribution function with mean zero and the correlation matrix:

$$\Sigma = \begin{pmatrix} I & \Sigma_2 \\ \Sigma_2 & I \end{pmatrix}$$

where :

$$\Sigma_2 = \begin{pmatrix} \rho & \rho.\sqrt{\frac{1-\rho_z}{1+\rho_z}} \\ \rho.\sqrt{\frac{1-\rho_w}{1+\rho_w}} & \rho.\sqrt{\frac{1-\rho_w}{1+\rho_w}}.\sqrt{\frac{1-\rho_z}{1+\rho_z}} \end{pmatrix}.$$

4.10 Simulation Procedures

4.10.1 *The general case*

To generate a sample (U_i, V_i), $i = 1, ..., n$ from a copula $C(u, v)$, one uses the fact that $C_u(v) = C(V/U = u)$ is a distribution function, and that $Z = C_u(V)$ obeys a uniform distribution on $[0, 1]$. Since U has a uniform distribution, its density on $[0, 1]$ is 1 and thus $C_u(v) = \frac{\partial C(u,v)}{\partial u}$. The procedure is as follows :

(1) First step :
Generate two random variables U and Z, independent and uniform on $[0, 1]$.
(2) Second step :
Calculate $V = C_u^{-1}(z)$. The pair (U, V) has the desired distribution. This procedure works well but it necessitates having an analytical expression for C_u^{-1}.

4.10.2 *Archimedean copulas*

In the case of Archimedian copulas, one can adapt the procedure described above. This method is due to Genest and Mackay [81]. Specifically: $\varphi(C) = \varphi(U) + \varphi(V)$ implies that

$$\varphi'(C)\frac{\partial C}{\partial u} = \varphi'(u).$$

An auxiliary variable $W = C(U, V)$ is calculated by :

$$W = (\varphi')^{-1}\left(\frac{\varphi'(u)}{\frac{\partial C}{\partial u}}\right) \tag{4.44}$$

where $(\varphi')^{-1}$ is the inverse function of the derivative of φ. The procedure is therefore as follows:

(1) First step:
Generate two uniform and independent random variables U and Z on $[0, 1]$.
(2) Second step :
Calculate W, using formula (4.44).
(3) Third step :

Calculate $V = \varphi^{-1}[\varphi(W) - \varphi(V)]$.

This procedure works well for the Clayton and Frank families. However for the Gumbel-Hougaard family we don't have an analytical expression for $(\varphi')^{-1}$.

For Gumbel-Hougaard family one can use the procedure developed by Lee [141]. Here one uses the fact that for any Archimedean copula, $T = \frac{\varphi(U)}{\varphi(U)+\varphi(V)}$ has a uniform distribution independent of $Z = C(U, V)$ (Section 4.6.6), and that for Gumbel-Hougaard family, $Z_1 = (\varphi(Z))^\alpha$ is distributed as a mixture of two gamma variates with the density $c(z) = (1 - \alpha + \alpha.z)exp(-z)$; here α the parameter of the copula. One then retrieves U and V from T and Z_1.

4.10.3 *Archimax distributions*

In this case the algorithm uses the same auxiliary variables $Z = C_{\varphi,A}(U, V)$ and $T = \frac{\varphi(U)}{\varphi(U)+\varphi(V)}$ as for Gumbel-Hougaard family described above. Specifically, the joint distribution function of (Z, T) is

$$P(Z \leq z, T \leq t) = K_\varphi(z)\left(t + t(1-t)\frac{A'(t)}{A(t)} + \lambda_\varphi(z)\int_0^t \frac{t(1-t)}{A(t)} dA'(t)\right)$$

where A' is the derivative of A, $K_\varphi(z) = z - \frac{\varphi(z)}{\varphi'(z)}$, is the distribution function of the variable Z and φ' stands for the right derivative wherever φ is not differentiable. In particular

$$P(T \leq t) \equiv H(t) = t + t(1-t)\frac{A'(t)}{A(t)} .$$

Suppose now that A possesses the second derivative A'' continuous everywhere on $(0, 1)$, and let $h(t)$ be the density of $H(t)$. Define

$$p(t) = \frac{t(1-t)A''(t)}{h(t)A(t)}, \quad 0 \leq t \leq 1$$

where it can be shown that $0 \leq p(t) \leq 1$ [85].

The conditional distribution of Z given $T = t$ is then:

$$P(Z \leq z/T = t) = p(t)z + (1 - p(t))K_\varphi(z) ,$$

This conditional distribution of Z given t is therefore a mixture of the univariate distribution K_φ and a uniform distribution on $(0, 1)$.

Thus the procedure is as follows :

(1) generate T from distribution H;
(2) given t the value of T, draw W from a uniform distribution on $[0,1]$;
(3) if the value w taken by W verifies $w \leq p(t)$, select Z from the uniform distribution on $[0,1]$, otherwise generate Z from distribution K_φ ;
(4) set $U = \varphi^{-1}\left(\frac{t.\varphi(z)}{A((t)}\right)$, and $V = \varphi^{-1}\left(\frac{(1-t).\varphi(z)}{A((t)}\right)$.

4.10.4 *Marshall and Olkin's mixture of distributions*

If the form of the functions K and G are simple (for example independent variables, upper Fréchet bound, etc.) one can use the following method:

(1) Generate $W = (W_1, ..., W_k)$ from G.
(2) Generate $V = (V_1, ..., V_n)$ from K.
(3) For $i = 1, ..., n$,
calculate $U_i = H_{i0}^{-1}(v_i^{1/w_i})$, where $H_{i0}(u) = \exp(-\varphi_i^{-1}(u))$, $\varphi_i^{-1}(t)$ being the Laplace transform of the marginal G_i, $i = 1, 2$ of G.

Details are provided in [152].

4.10.5 *Three-dimensional copulas with truncation invariance*

In this case one uses the particular structure of $C_{123}(u, v, w)$:

$$C(u,v,w) = C_{12}\left(\frac{C_{13}(u,w)}{w}, \frac{C_{23}(v,w)}{w}\right)w.$$

(1) Generate a uniform variable W on $[0,1]$
(2) Generate the pair (X,Y) from the C_{12} distribution
(3) From $Z_1 = WX$ and $Z_2 = WY$, determine $U = C_{13/W=w}^{-1}(z_1)$ and $V = C_{23/W=w}^{-1}(z_2)$.

An interested reader is referred to [40] for additional details.

Chapter 5

Farlie-Gumbel-Morgenstern Models of Dependence

5.1 Introduction

In this chapter, we shall summarize in some detail statistical and probabilistic properties of a popular well-known family (which seems to be quite natural –at least empirically) of bivariate dependent variables and its numerous generalizations scattered in the literature. Dependence properties of this family are closely associated with the correlation coefficient although *a priori* the pivotal parameter of the family is not obviously associated with this concept. Some of the results presented in this chapter appear for the first time in monographic (or periodic) form. In the author's opinion the family and its generalizations are ideally suited for various models with small or moderate dependence and do not depend on a particular physical model which may or may not be appropriate in a given situation.

It should be emphasized that all the distributions discussed herein, can be viewed as particular case of the multivariate distributions obtained using the construction method developed by Rüschendorf in his pioneering paper 1985 discussed in some details in Chapter 4. In fact, it was Rüschendorf who inspired –albeit indirectly– investigations of many a distribution developed by the authors of this monograph.

Order statistics from Farlie-Gumbel-Morgenstern (FGM) distributions seems to be especially attractive. Some novel results on this topic are presented in this chapter.

5.2 Initial Definition

Morgenstern (1956), Farlie (1960) and Gumbel (1958) have discussed families of bivariate distributions of the form

$$F_{\mathbf{X}}(\mathbf{x}) = F_{X_1,X_2}(x_1, x_2)$$

$$= F_{12}(x_1, x_2) = F_1(x_1)F_2(x_2)\left[1 + \alpha S_1(x_1)S_2(x_2)\right] \quad (|\alpha| < 1) \tag{5.1}$$

where $F_{\mathbf{X}}(\mathbf{x})$ is the joint cumulative distribution function of X_1 and X_2, $F_j(x_j) = F_{X_j}(x_j)$ and $S_j(x_j) = 1 - F_j(x_j)$; $j = 1, 2$.

Equation (5.1) is consistent in the sense $F_{X_1,X_2}(x_1, \infty) = F_1(x_1)$, etc.

If the densities $f_{\mathbf{X}}(.)$ (corresponding to $F_{\mathbf{X}}(.)$) exist then (5.1) implies :

$$f_{\mathbf{X}}(\mathbf{x}) = f_{12}(x_1, x_2) = f_1(x_1)f_2(x_2)\left[1 + \alpha\left\{1 - 2F_1(x_1)\right\}\left\{1 - 2F_2(x_2)\right\}\right]. \tag{5.2}$$

For the densities $f_{\mathbf{X}}(\mathbf{x}), f_1(x_1)$ and $f_2(x_2)$ to exist we must have $|\alpha| < 1$. Generally, however,

$$-\frac{1}{\max\left\{(1-m_1)(1-m_2), M_1M_2\right\}} \leq \alpha \leq \frac{1}{\max\left\{(1-m_1)(1-m_2), M_1M_2\right\}},$$

where $m_j = \inf H_j$; $M_j = \sup H_j$; $H_j = \{F_j(x) : -\infty < x < \infty\} \bigcap \{0, 1\}$, $j = 1, 2$.

We note that if (X_1, X_2) have a joint FGM distribution and $Y_1 = h_1(X_1)$, $Y_2 = h_2(X_2)$ are monotonic increasing functions of X_1, X_2, respectively, then Y_1 and Y_2 also possess a joint FGM distribution. This is easily seen by noting that

$$P\left\{\bigcap_{j=1}^{2}(Y_j \leq h_j(x_j))\right\} = F_{12}(x_1, x_2)$$

and

$$P\{Y_j \leq h_j(x_j)\} = F_j(x_j).$$

If X_1 and X_2 are each continuous, we can find transformations $Y_1 = h_1(X_1)$, $Y_2 = h_2(X_2)$ so that Y_1 and Y_2 each have a standard uniform distribution. The resulting special FGM– a copula– distribution (with the uniform marginals) was discussed in 1936 by H. Eyraud, which is, we believe, the earliest reference to FGM distributions. (See also Kimeldorf and Sampson (1975) [127].)

In terms of survival functions

$$S_{12}(x_1, x_2) = P\left\{\bigcap_{j=1}^{2} (X_j > x_j)\right\}; \; S_j(x_j) = P\{X_j > x_j\},$$

(5.1) is equivalent to

$$S_{12}(x_1, x_2) = S_1(x_1)S_2(x_2)\{1 + \alpha F_1(x_1)F_2(x_2)\}. \tag{5.3}$$

Note the symmetry with the definition (5.1).

Further investigations are due to Johnson and Kotz (1975, 1977) and Cambanis (1977).

5.3 Regression and Correlation

It follows from straightforward calculations that

$$E\{X_2 \mid X_1\} = E\{X_2\} + \alpha J_2 \{2F_1(X_1) - 1\},$$

where

$$J_2 = \int_{-\infty}^{\infty} F_2(x)(1 - F_2(x))\,dx$$

and

$$J_1 = \int_{-\infty}^{\infty} F_1(x)(1 - F_1(x))\,dx$$

$$\lim_{X_1 \to -\infty} E\{X_2 \mid X_1\} = E\{X_2\} - \alpha J_2, \; \lim_{X_1 \to \infty} E\{X_2 \mid X_1\} = E\{X_2\} + \alpha J_2,$$

$$cov(X_1, X_2) = \alpha J_1 J_2,$$

and

$$var(X_2 \mid X_1) = var(X_2) + 2\alpha\{2F_1(X_1) - 1\}\{K_2 - J_2 E[X_2]\}$$
$$-\alpha^2 J_2^2 \{2F_1(X_1) - 1\}$$

with

$$K_2 = \int_{-\infty}^{\infty} x F_2(x) \left[1 - F_2(x)\right] dx,$$

$$\lim_{X_1 \to -\infty} var(X_2 \mid X_1) = var\left\{X_2\right\} - 2\alpha\left\{K_2 - J_2 E\left[X_2\right]\right\} - \alpha^2 J_2^2,$$

$$\lim_{X_1 \to \infty} var(X_2 \mid X_1) = var\left\{X_2\right\} + 2\alpha\left\{K_2 - J_2 E\left[X_2\right]\right\} - \alpha^2 J_2^2.$$

If $var(X_j) = 1$ and $E\left\{X_j\right\} = 0$ (standardized variables) then

$$corr(X_1, X_2) = \alpha J_1 J_2.$$

If also X_2 has a symmetric distribution, then $K_2 = 0$ and

$$var(X_2 \mid X_1) = var(X_2) - \alpha^2 J_2^2 \left\{2F_1(X_1) - 1\right\}^2.$$

The cases $\alpha = 1$ and $\alpha = -1$ represent the maximal degrees of positive and negative dependence, respectively, allowed in this family.

a) If the marginals are $N(0,1)$ the correlation is $\alpha\pi^{-1}$ (ranges from -0.318 to 0.318).

b) If the marginals are uniform distributions over (0,1), the correlation is $\alpha/3$ (so it ranges from $-\frac{1}{3}$ to $-\frac{1}{3}$). In fact, for FGM distributions with absolutely continuous marginals, the correlation coefficient between X_1 and X_2 cannot exceed $\frac{1}{3}$(Schucany *et al.* [194]).

The FGM model can also be expressed as follows:

For $\alpha \in [0,1]$:

$$f_{12}(x_1, x_2) = (1-\alpha) f_1(x_1) f_2(x_2) + \alpha f_1(x_1) f_2(x_2) \left\{1 + \left[1 - 2F_1(x_1)\right]\left[1 - 2F_2(\right.\right.$$

and for $\alpha \in [-1, 0]$:

$$f_{12}(x_1, x_2) = (1+\alpha) f_1(x_1) f_2(x_2) - \alpha f_1(x_1) f_2(x_2) \left\{1 - \left[1 - 2F_1(x_1)\right]\left[1 - 2F_2(\right.\right.$$

$$= (1-\mu) f_1(x_1) f_2(x_2) + \mu f_1(x_1) f_2(x_2) \left\{1 - \left[1 - 2F_1(x_1)\right]\left[1 - 2F_2(x_2)\right]\right\},$$

where $\mu = -\alpha \in [0,1]$.

This representation naturally suggests the following generalization. Define the class of densities (De la Horra and Fernandez (1995)):

$$\Gamma = \Gamma^+ \cup \Gamma^-,$$

where

$$\Gamma^+ = \left\{f(x_1,x_2) = (1-\alpha)f_I(x_1,x_2) + \alpha f^+(x_1,x_2), \alpha \in [0,1]\right\},$$

$$\Gamma^- = \left\{f(x_1,x_2) = (1-\alpha)f_I(x_1,x_2) + \alpha f^-(x_1,x_2), \alpha \in [0,1]\right\},$$

$f_I(x_1x_2) = f_1(x_1)f_2(x_2)$ is the density obtained under independence and $f^+(x_1,x_2)$ and $f^-(x_1,x_2)$ are fixed densities with marginals $f_1(x_1)$ and $f_2(x_2)$, representing some degree of positive and negative dependence, respectively. Obviously, the FGM model is a special case of this family, with

$$f^+(x_1,x_2) = f_1(x_1)f_2(x_2)\left\{1 + [1-2F_1(x_1)]\,[1-2F_2(x_2)]\right\}$$

and

$$f^-(x_1,x_2) = f_1(x_1)f_2(x_2)\left\{1 - [1-2F_1(x_1)]\,[1-2F_2(x_2)]\right\}.$$

The class Γ^+ is quite similar to the class of distributions considered in De la Horra and Ruiz-Rivas (1988).

However

$$F_{12}(x_1,x_2) = F_1(x_1)F_2(x_2)[1 + \alpha S_1(x_1^{\varphi_1})S_2(x_2^{\varphi_2})] \;\; 0 < \varphi_j < 1,\; j = 1,2,$$

is not a proper distribution with absolutely continuous marginals. Indeed its density has negative values for some pairs (x_1,x_2).

5.4 Iterations

From

$$F(x_1,x_2) = F_1(x_1)F_2(x_2)\,[1 + \alpha S_1(x_1)S_2(x_2)]$$

one can construct the following generalization

$$F = F_1F_2\left\{1 + \alpha_1 S_1S_2 + \alpha_2 F_1F_2S_1S_2 + \ldots + \alpha_k\,(F_1F_2)^{[\frac{k}{2}]}\,(S_1S_2)^{[\frac{k}{2}+\frac{1}{2}]}\right\}$$

$$= F_1F_2 + \sum_{j=1}^{k} \alpha_j\,(F_1F_2)^{[\frac{j}{2}]+1}\,(S_1S_2)^{[\frac{j}{2}+\frac{1}{2}]}. \tag{5.4}$$

Evidently, F still has the marginals F_1 and F_2. We shall call (5.4) the "$k-1$ fold iteration". In particular, the bivariate FGM with a single iteration ($k=2$) can be written as

$$F = F_1F_2 + \alpha F_1F_2S_1S_2 + \beta\left(F_1F_2\right)^2 S_1S_2. \tag{5.5}$$

Examples of correlation values for the distribution (5.5) are :

For Uniform marginals:

$$corr(X,Y) = \frac{\alpha}{3} + \frac{\beta}{12}.$$

For Normal marginals

$$corr(X,Y) = \frac{\alpha}{\pi} + \frac{\beta}{4\pi}.$$

The ranges of parameters for iterated FGM with absolutely continuous marginals are :

$$|\alpha| \le 1,\ |\alpha+\beta| \le -1,\ \beta \le 2^{-1}\left\{3-\alpha+(9-6\alpha-3\alpha^2)^{\frac{1}{2}}\right\}$$

(and not $|\beta| \le 1$ as it was initially assumed). In this case $\rho = corr(X,Y)$ yields maximal value $6^{-1}(13^{\frac{1}{2}}-1) = 0.43426$ for uniform marginals and $(2\pi)^{-1}\left(13^{\frac{1}{2}}-1\right) = 0.41469$ for normal marginals. Equivalently, for uniform marginals

$$F(x_1,x_2) = x_1x_2\left\{1+\alpha(1-x_1)(1-x_2)+\beta x_1x_2(1-x_1)(1-x_2)\right\},$$

$$0 < x, y < 1$$

and

$$f(x_1,x_2) = 1+\alpha(1-2x_1)(1-2x_2)+\beta x_1x_2(2-3x_1)(2-3x_2),\ 0 \le x, y \le 1.$$

Zheng and Klein (1999) study the FGM copula :

$$C(u,v) = uv(1+\alpha(1-u)(1-v)),\ -1 \le \alpha \le 1.$$

which is a polynomial not Archimedean copula (as indicated in Chapter 4) and its iteration

$$C(u,v) = uv + \Sigma_j\alpha_j(uv)^{1/2}[(1-u)(1-v)]^{(j+1)/2},\ -1 \le \alpha_j \le 1.$$

They are of particular interest for self-consistent estimators of survival function $\hat{S}(t)$ and the reliability $\hat{R}(t)$. See Huang and Kotz (1984) for more details.

5.5 Dependence Properties

The FGM distribution is LRD (its density is TP_2) for $0 \leq \alpha \leq 1$, hence it is also PRD in X and Y, associated, RTI, LTD and PQD. Its density is RR_2 if $-1 \leq \alpha \leq 0$.

Indeed given $f(x,y) = f(x)f(y)\left[1 + \alpha\{1 - 2F_1(x)\}\{1 - 2F_2(y)\}\right]$ for $x_1 < x_2$ and $y_1 < y_2$, we have :

$$\begin{aligned} & f(x_1,y_1)f(x_2,y_2) - f(x_1,y_2)f(x_2,y_1) \\ = \;& 4\alpha f(x_1)f(x_2)f(y_1)f(y_2)(F(x_2) - F(x_1))(F(y_2) - F(y_1)) \end{aligned}$$

which is clearly ≥ 0 for $0 \leq \alpha \leq 1$ and negative if α is negative.

5.6 A Class of n-variate FGM Distributions

We generalize the FGM system of distributions with specified marginal distributions for the n-variate case using the following formula for a cumulative distribution function:

$$F(x_1, ..., x_n) = \left[1 + \sum_{s=2}^{n} \sum_{1 \leq i_1 < ... < i_s \leq n} \alpha_{i_1,...,i_s} \prod_{j=1}^{s} S_{i_j}(x_{i_j})\right] \prod_{j=1}^{n} F_j(x_j), \tag{5.6}$$

where the total number of parameters $\alpha_{i_1,...,i_s}$ is $card_{2 \leq j \leq n}\{\alpha_{i_1,...,i_s}\} = 2^n - n - 1$. More explicitly:

$$\begin{aligned} F_{\mathbf{X}}(\mathbf{x}) = \prod_{j=1}^{n} F_j(x_j)[1 &+ \sum_{j_1<j_2} \alpha_{j_1 j_2} S_{j_1}(x_1) S_{j_2}(x_2) \\ &+ \sum_{j_1<j_2<j_3} \alpha_{j_1 j_2 j_3} S_{j_1}(x_1) S_{j_2}(x_2) S_{j_3}(x_3) +, ..., \end{aligned}$$

$$+\alpha_{12\ldots n}\prod_{j=1}^{n}S_j(x_j)]. \tag{5.7}$$

As an example, an explicit form for the three-dimensional FGM system is :

$$\begin{aligned}F(x_1,x_2,x_3) &= F_1(x_1)F_2(x_2)F_3(x_3)(1+\alpha_{12}(1-F_1(x_1))(1-F_2(x_2))\\ &+\alpha_{13}(1-F_1(x_1))(1-F_3(x_3))+\alpha_{23}(1-F_2(x_2))(1-F_3(x_3))\\ &+\alpha_{123}(1-F_1(x_1))(1-F_2(x_2))(1-F_3(x_3))).\end{aligned} \tag{5.8}$$

More details are presented in Section 5.6.1.

$F_j(x_j)$ denotes the cumulative distribution function of the specified marginal distributions, $j = 1, 2, ..., n$. The probability density functions of this generalized FGM family may be written as

$$f(x_1,\ldots,x_n) = \left[1+\sum_{s=2}^{n}\sum_{1\le i_1<\ldots<i_s\le n}\alpha_{i_1,\ldots,i_s}\prod_{j=1}^{s}(1-2F_{i_j}(x_{i_j}))\right]\prod_{j=1}^{n}f_j(x_j)\,. \tag{5.9}$$

The marginal distributions related to distributions given by (5.9) have probability density functions of a similar form

$$f(x_{k_1},\ldots,x_{k_m}) = \left[1+\sum_{s=2}^{n}\sum_{k_1\le r_1<\ldots<r_s\le k_m}\alpha_{r_1,\ldots,r_s}\prod_{j=1}^{s}(1-2F_{r_j}(x_{r_j}))\right]$$

$$\times\prod_{j=1}^{n}f_{k_j}(x_{k_j})$$

$$1\le k_j<k_{j+1}\le n.$$

It is easy to observe that parameters in (5.6)–(5.9) must satisfy the following conditions

$$1+\sum_{s=2}^{m}\sum_{k_1\le r_1<\ldots<r_s\le k_m}\alpha_{r_1,\ldots,r_s}\prod_{j=1}^{n}\zeta_{r_j}\ge 0 \qquad where \qquad |\zeta_{r_j}|\le 1$$

for $1 \le k_1 < k_m \le n$ and successively for $m = 2, 3, ..., n$.

The conditional distributions connected with distributions (5.9) have conditional density functions of form:

$$\begin{aligned} &f(x_{k_1},...,x_{k_i} \mid x_{k_{i+1}},...,x_{k_n}) \\ &= \left\{1+\left[1+\sum_{s=2}^{n-i}\sum_{k_{i+1}\le h_1<\ldots<h_s\le k_n}\alpha_{h_1,\ldots,h_s}\prod_{j=1}^{s}(1-2F_{h_j}(x_{h_j}))\right]^{-1}\right. \\ &\qquad \times\left.\left[\sum_{s=2}^{n}\sum_{G}\alpha_{r_1,\ldots,r_s}\prod_{j=1}^{s}(1-2F_{r_j}(x_{r_j}))\right]\right\}\prod_{j=1}^{i}f_{k_j}(x_{k_j}) \qquad (5.10) \end{aligned}$$

where $G = \{k_1 \le r_1 < ... < r_s \le k_n, (r_1, r_2, ..., r_s) \ne (h_1, h_2, ..., h_s)\}$ with $s = 2, ..., n - i$.

The conditional distributions for the FGM family are conveniently defined by the survival functions :

$$S_1\left(x_1/\bigcap_{1}^{m}X_j>x_j\right) = \frac{S_1(x_1)\left(1+\sum_{j<j_1}\alpha_{jj_1}F_j(x_j)F_{j_1}(x_{j_1})-\sum_{j<j_1<j_2}\alpha_{jj_1j_2}F_j(x_j)F_{j_1}(x_{j_1})F_{j_2}(x_{j_2})\right)}{D}$$

where

$$\begin{aligned} D = 1 &+ \sum_{j_1<j_2}\alpha_{j_1j_2}(1-2F_{j_1}(x_{j_1}))(1-2F_{j_2}(x_{j_2})) + \ldots \\ &+\alpha_{12\ldots n}\prod_{j=1}^{n}(1-2F_j(x_j)). \end{aligned}$$

5.6.1 *A class of bivariate FGM distributions with Weibull marginal distributions*

Bivariate FGM with exponential marginals is well known as bivariate Gumbel exponential distribution, widely discussed in the literature. Here we shall study its extension for Weibull marginals. This distribution was also described by Johnson and Kotz (1975) and Lee (1977).

By substituting

$$F_i(x_i) = 1 - \exp\left(-\left(\frac{x_i}{b_i}\right)^{\beta_i}\right), \qquad i = 1, 2$$

into Eq.(5.1), we obtain a class of bivariate distributions with Weibull marginals

$$F(x_1, x_2) = \left\{1 - \exp\left(-\left(\frac{x_1}{b_1}\right)^{\beta_1}\right)\right\}\left\{1 - \exp\left(-\left(\frac{x_2}{b_2}\right)^{\beta_2}\right)\right\} \times \left\{1 + \alpha \exp\left(-\left(\frac{x_1}{b_1}\right)^{\beta_1} - \left(\frac{x_2}{b_2}\right)^{\beta_2}\right)\right\} \tag{5.11}$$

where $x_i > 0, \beta_i > 0$ (a curve shape parameter), $b_i > 0$ (a scale parameter), $i = 1, 2$ and a dependence parameter $\alpha \in (-1, 1)$.

From Eq.(5.11) it follows that the density functions of a class of bivariate distributions with Weibull marginal distributions are of form :

$$f(x_1, x_2) = \frac{\beta_1}{b_1}\frac{\beta_2}{b_2}\left(\frac{x_1}{b_1}\right)^{\beta_1 - 1}\left(\frac{x_2}{b_2}\right)^{\beta_2 - 1}\left\{1 + \alpha\left(1 - 2\exp\left(-\left(\frac{x_1}{b_1}\right)^{\beta_1}\right)\right) \times \left(1 - 2\exp\left(-\left(\frac{x_2}{b_2}\right)^{\beta_2}\right)\right)\right\}\exp\left(-\left(\frac{x_1}{b_1}\right)^{\beta_1} - \left(\frac{x_2}{b_2}\right)^{\beta_2}\right) \tag{5.12}$$

where $x_i > 0$, $\beta_i > 0$, $b_i > 0$; $i = 1, 2$, $\alpha \in (-1, 1)$.

The distribution functions (5.11) and the density functions (5.12) are expressed in the forms which is symmetric in the variables, it is therefore sufficient to consider a conditional distribution of X_i, given $X_j = x_j$, $i \neq j$,

$i = 1, 2$, when the joint density of X_1, X_2 is described by (5.12). Hence the conditional density function of X_i, given $X_j = x_j$, $i \neq j$, $i = 1, 2$ is

$$f(x_i \mid x_j) = \left\{1 + \alpha\left(1 - 2\exp\left(-\left(\frac{x_1}{b_1}\right)^{\beta_1}\right)\right)\left(1 - 2\exp\left(-\left(\frac{x_2}{b_2}\right)^{\beta_2}\right)\right)\right\}$$

$$\times \frac{\beta_i}{b_i}\left(\frac{x_i}{b_i}\right)^{\beta_i - 1} \exp\left(-\left(\frac{x_i}{b_i}\right)^{\beta_i}\right).$$

The regression of X_i on X_j is given by

$$E\left(X_i \mid X_j = x_j\right) = b_i \Gamma(1 + \frac{1}{\beta_i})$$

$$\times \left\{1 + \alpha\left(1 - 2\exp\left(-\left(\frac{x_j}{b_j}\right)^{\beta_j}\right)\right)(1 - 2^{-1/\beta_i})\right\}, \qquad (5.13)$$

$i \neq j$; $i, j = 1, 2$. The conditional expectation $E\left(X_i \mid X_j = x_j\right)$ increases if $\alpha > 0$ (decreases if $\alpha < 0$) starting from $b_i \Gamma(1 + \frac{1}{\beta_i})\left[1 - \alpha(1 - 2^{-1/\beta_i})\right]$ to $b_i \Gamma(1 + \frac{1}{\beta_i})\left[1 + \alpha(1 - 2^{-1/\beta_i})\right]$ provided that the variable X_j increases. From (5.13), denoting $E\left(X_i \mid X_j = x_j\right) = E_\alpha\left(X_i \mid x_j\right)$, it is easy to obtain the following relation :

$$\frac{1}{2}\left[E_{-\alpha}\left(X_i \mid x_j\right) + E_\alpha\left(X_i \mid x_j\right)\right] = b_i \Gamma(1 + \frac{1}{\beta_i}) \quad i \neq j, i, j = 1, 2.$$

This relationship indicates the symmetry between $E_\alpha\left(X_i \mid x_j\right)$ and $E_{-\alpha}\left(X_i \mid x_j\right)$. Conditional moments of order k possess a similar property :

$$\frac{1}{2}\left[E_{-\alpha}\left(X_i^k \mid x_j\right) + E_\alpha\left(X_i^k \mid x_j\right)\right] = b_i^k \Gamma(1 + \frac{k}{\beta_i}).$$

The conditional variance of X_i , given $X_j = x_j$, is given by

$$Var\left(X_i \mid x_j\right) = b_i^2 \Gamma(1 + \frac{2}{\beta_i})\left\{1 + \alpha\left(1 - 2\exp\left(-\left(\frac{x_j}{b_j}\right)^{\beta_j}\right)\right)(1 - 2^{-2/\beta_i})\right\}$$

$$-b_i^2\Gamma^2(1+\frac{1}{\beta_i})\left\{1+\alpha\left(1-2\exp\left(-\left(\frac{x_j}{b_j}\right)^{\beta_j}\right)\right)(1-2^{-1/\beta_i})\right\}^2$$

$$i \neq j, i, j = 1, 2.$$

It is easy to verify that

$$E(X_1^{u_1}X_2^{u_2}) = b_1^{u_1}b_2^{u_2}\Gamma(1+\frac{u_1}{\beta_1})\Gamma(1+\frac{u_2}{\beta_2})\left[1+\alpha(1-2^{-u_1/\beta_1})(1-2^{-u_2/\beta_2})\right]$$

$(u_1, u_2) \in R_+^2$.

In particular, using this expression the covariance between X_1 and X_2 is found to be

$$Cov(X_1, X_2) = \alpha b_1 b_2 \Gamma(1+\frac{1}{\beta_1})\Gamma(1+\frac{1}{\beta_2})(1-2^{-1/\beta_i})(1-2^{-1/\beta_2})$$

and the correlation coefficient is

$$\rho = \alpha\frac{(1-2^{-1/\beta_1})(1-2^{-1/\beta_2})\Gamma(1+\frac{1}{\beta_1})\Gamma(1+\frac{1}{\beta_2})}{\left(\left[\Gamma(1+\frac{2}{\beta_1})-\Gamma^2(1+\frac{1}{\beta_1})\right]\left[\Gamma(1+\frac{2}{\beta_2})-\Gamma^2(1+\frac{1}{\beta_2})\right]\right)^{1/2}}.$$

Note that the least squares linear regression of X_i on X_j has as its equation

$$E(X_i \mid X_j) = \alpha\frac{b_j\Gamma(1+\frac{1}{\beta_1})\Gamma(1+\frac{1}{\beta_2})(1-2^{-1/\beta_1})(1-2^{-1/\beta_2})}{b_i\left[\Gamma(1+\frac{2}{\beta_i})-\Gamma^2(1+\frac{1}{\beta_i})\right]}X_j$$

$$+b_j\Gamma(1+\frac{2}{\beta_i})\left\{1-\alpha\frac{\Gamma^2(1+\frac{1}{\beta_i})(1-2^{-1/\beta_1})(1-2^{-1/\beta_2})}{\left[\Gamma(1+\frac{2}{\beta_i})-\Gamma^2(1+\frac{1}{\beta_i})\right]}\right\},$$

a linear function in X_j.

5.6.2 *A class of three-variate distributions with Weibull marginal distributions*

As it was mentioned above, the FGM system of bivariate distributions with specified marginal distributions may be extended in case of three random variables as follows

$$F(x_1, x_2, x_3) = \left[1 + \sum_{1 \le r < s \le 3} \alpha_{rs} S_r(x_r) S_s(x_s) + \alpha_{123} \prod_{j=1}^{3} S_j(x_j)\right] \prod_{j=1}^{3} F_j(x_j) \tag{5.14}$$

where $S_i(x_i) = 1 - F_i(x_i)$ and $F_i(x_i)$ is a specified cumulative distribution function, $i = 1, 2, 3$, and the parameters $\alpha_{rs} \in (-1, 1)$, $1 \le r < s \le 3$.

The density functions describing the system of three-variate distributions with specified marginal distributions $F_i(x_i)$, $i = 1, 2, 3$ are of form

$$\begin{aligned} f(x_1, x_2, x_3) &= \left[1 + \sum_{1 \le r < s \le 3} \alpha_{rs} (1 - 2F_r(x_r))(1 - 2F_s(x_s)) \right. \\ &+ \left. \alpha_{123} \prod_{j=1}^{3} (1 - 2F_j(x_j))\right] \times \prod_{j=1}^{3} f_j(x_j) \, . \end{aligned} \tag{5.15}$$

The marginal bivariate distributions associated with distributions having cumulative distribution functions (5.14) or probability density functions (5.15) are given by

$$F_{ij}(x_i, x_j) = F_i(x_i) F_j(x_j) \left[1 + \alpha_{ij} S_i(x_i) S_j(x_j)\right] \quad , 1 \le i < j \le 3 \, ,$$

or

$$f_{ij}(x_i, x_j) = f_i(x_i) f_j(x_j) \left[1 + \alpha_{ij} (1 - 2F_i(x_i))(1 - 2F_j(x_j))\right] \quad , 1 \le i < j \le 3 \, .$$

A trivial case of independent random variables implies $\alpha_{ij} = 0$, $1 \le i < j \le 3$, $\alpha_{123} = 0$. The conditional distributions connected with distributions (5.15) may be described by conditional probability density functions

$$
\begin{aligned}
f(x_i \mid x_j, x_k) &= f_i(x_i)\{1 + [1 + \alpha_{jk}(1 - 2F_j(x_j))(1 - 2F_k(x_k))]^{-1} \\
&\times [(1 - 2F_i(x_i))\,(\alpha_{ij}(1 - 2F_i(x_i)) + \alpha_{ik}(1 - 2F_k(x_k))) \\
&+ \alpha_{123} \prod_{r=1}^{3} (1 - 2F_r(x_r))]\}
\end{aligned}
$$

where $i \neq j$, $i \neq k, j < k; i, j, k \in \{1, 2, 3\}$.

$$
f(x_i, x_j \mid x_k)
$$

$$
= f_i(x_i) f_j(x_j) \left[1 + \sum_{1 \leq r < s \leq 3} \alpha_{rs}(1 - 2F_r(x_r))(1 - 2F_s(x_s)) \right.
$$

$$
\left. + \alpha_{123} \prod_{r=1}^{3} (1 - 2F_r(x_r)) \right]
$$

where $j > i$, $i \neq k$, $j \neq k$; $i, j, k \in \{1, 2, 3\}$.

Using formulas (5.14)–(5.15) to obtain a system of three-variate distributions with Weibull marginal distributions we set

$$
F_i(x_i) = 1 - \exp(-x_i^{\beta_i}) \quad \textit{or} \quad f_i(x_i) = \beta_i x_i^{\beta_i - 1} \exp(-x_i^{\beta_i}) \;\; x_i > 0, \beta_i > 0. \tag{5.16}
$$

Here it is assumed without loss of generality, that $b_i = 1$, $i = 1, 2, 3$, for the cumulative distribution functions and probability distribution functions of a class of three-variate distributions with Weibull marginal distributions

$$
F(x_1, x_2, x_3) = \left[1 + \sum_{1 \leq r < s \leq 3} \alpha_{rs} \exp(-x_r^{\beta_r} - x_s^{\beta_s}) + \alpha_{123} \exp(-\sum_{r=1}^{3} x_r^{\beta_r}) \right]
$$

$$
\times \prod_{j=1}^{3} (1 - \exp(-x_j^{\beta_j})) \tag{5.17}
$$

$$\begin{aligned} f(x_1, x_2, x_3) &= \left[1 + \sum_{1 \leq r < s \leq 3} \alpha_{rs}(1 - 2\exp(-x_r^{\beta_r}))(1 - 2\exp(-x_s^{\beta_s})) \right. \\ &\quad \left. -\alpha_{123} \prod_{j=1}^{3}(1 - 2\exp(-x_j^{\beta_j}))\right] \\ &\quad \times \prod_{j=1}^{3} \beta_j x_j^{\beta_j - 1} \exp(-x_j^{\beta_j}) \;\; \alpha_{rs} \in (-1, 1) \end{aligned}$$

$$(x_1, x_2, x_3) \in R_+^3. \tag{5.18}$$

The conditional probability density functions for this class of distributions (5.18) are

$$\begin{aligned} f(x_i \mid x_j, x_k) &= \{1 + [(1 - 2\exp(-x_i^{\beta_i}))(\alpha_{ij}(1 - 2\exp(-x_j^{\beta_j})) \\ &\quad +\alpha_{ik}(1 - 2\exp(-x_k^{\beta_k})) - \alpha_{123} \prod_{r=1}^{3}(1 - 2\exp(-x_r^{\beta_r}))] \\ &\quad \left[1 + \alpha_{jk}(1 - 2\exp(-x_j^{\beta_j})) + (1 - 2\exp(-x_k^{\beta_k}))\right]^{-1}\} \\ &\quad \times \beta_i x_i^{\beta_i - 1} \exp(-x_i^{\beta_i}) \end{aligned}$$

where $i \neq j$, $i \neq k$; $j < k; i, j, k \in \{1, 2, 3\}$.

$$f(x_i, x_j \mid x_k)$$

$$= [1 + \sum_{1 \leq r < s \leq 3} \alpha_{rs}(1 - 2\exp(-x_r^{\beta_r}))(1 - 2\exp(-x_s^{\beta_s}))$$

$$-\alpha_{123} \prod_{r=1}^{3}(1 - 2\exp(-x_r^{\beta_r}))] \prod_{r \neq k}^{3} \beta_r x_r^{\beta_r - 1} \exp(-x_r^{\beta_r})$$

where $i < j$, $i \neq k$; $j \neq k; i, j, k \in \{1, 2, 3\}$.

The regression of X_i on X_j, X_k is

$$E(X_i \mid x_j, x_k) = \{1 + [1 + \alpha_{jk}(1 - 2\exp(-x_j^{\beta_j})) + (1 - 2\exp(-x_k^{\beta_k}))]^{-1} \times (1 - 2^{-1/\beta_i})[(\alpha_{ij}(1 - 2\exp(-x_j^{\beta_j})) + \alpha_{ik}(1 - 2\exp(-x_k^{\beta_k})) - \alpha_{123}\prod_{r=1}^{3}(1 - 2\exp(-x_r^{\beta_r}))]\}\Gamma(1 + \frac{1}{\beta_i})$$

$i \neq j,\ i \neq k\ ;\ j < k; i, j, k \in \{1, 2, 3\}$.

Taking into account the expression

$$\begin{aligned} E(X_1^{u_1} X_2^{u_2} X_3^{u_3}) &= [1 + \sum_{1 \le i < j \le 3} \alpha_{ij}(1 - 2^{-u_i/\beta_i})(1 - 2^{-u_j/\beta_j}) \\ &- \alpha_{123}\prod_{r=1}^{3}(1 - 2^{-u_r/\beta_r})]\prod_{r=1}^{3}\Gamma(1 + \frac{u_r}{\beta_r}) \\ &(u_1, u_2, u_3) \in R_+^3. \end{aligned}$$

One easily obtains the elements of the covariance matrix $\{Cov(X_i, X_j)\}$; $(i, j = 1, 2, 3)$. Namely

$$Cov(X_i, X_j) = \begin{cases} \alpha_{ij}\Gamma(1 + \frac{1}{\beta_i})\Gamma(1 + \frac{1}{\beta_j})(1 - 2^{-1/\beta_i})(1 - 2^{-1/\beta_j}) & i \neq j \\ \Gamma(1 + \frac{2}{\beta_i}) - \Gamma^2(1 + \frac{1}{\beta_i}) & i = j \end{cases}. \tag{5.19}$$

It is easy to verify that $Det\{Cov(X_i, X_j)\} \neq 0$. The entries of the correlation matrix $\{\rho_{ij}\}$ are of form

$$\rho_{ij} = \begin{cases} \alpha_{ij}\dfrac{(1-2^{-1/\beta_i})(1-2^{-1/\beta_j})\Gamma(1+\frac{1}{\beta_i})\Gamma(1+\frac{1}{\beta_j})}{\left(\left[\Gamma(1+\frac{2}{\beta_i})-\Gamma^2(1+\frac{1}{\beta_i})\right]\left[\Gamma(1+\frac{2}{\beta_j})-\Gamma^2(1+\frac{1}{\beta_j})\right]\right)^{1/2}} & i \neq j \\ 1 & i = j \end{cases}. \tag{5.20}$$

Numerical calculations for the values of ρ_{ij} are straightforward.

5.6.3 *FGM n-variate distributions with Weibull marginals*

The construction of a class of n–variable distributions with Weibull marginal distributions is carried out by inserting (5.16) into (5.6) or (5.9). Hence this class of n–variable distributions with Weibull marginal distributions is described by cumulative distribution functions

$$F(x_1, ..., x_n) = \left[1 + \sum_{s=2}^{n} \sum_{1 \le i_1 < ... < i_s \le n} \alpha_{i_1,...,i_s} \prod_{j=1}^{s} \exp(-x_{i_j}^{\beta_{i_j}})\right] \prod_{j=1}^{n}(1 - \exp(-x_j^{\beta_j})) \tag{5.21}$$

or by probability density functions

$$f(x_1, ..., x_n) = \left[1 + \sum_{s=2}^{n} \sum_{1 \le i_1 < ... < i_s \le n} (-1)^s \alpha_{i_1,...,i_s} \prod_{j=1}^{s}(1 - 2\exp(-x_{i_j}^{\beta_{i_j}}))\right] \times \prod_{j=1}^{n} \beta_j x_j^{\beta_j - 1}(1 - \exp(-x_j^{\beta_j})). \tag{5.22}$$

The product moments of $X_1, X_2, ..., X_n$ possessing the distribution (5.21) can be obtained by using the functions

$$E(X_1^{u_1} ... X_n^{u_n}) = \left[1 + \sum_{s=2}^{n} \sum_{1 \le i_1 < ... < i_s \le n} (-1)^s \alpha_{i_1,...,i_s} \prod_{j=1}^{s}(1 - 2^{-u_j/\beta_j})\right] \times \prod_{j=1}^{n} \Gamma\left(1 + \frac{u_j}{\beta_j}\right), \quad (u_1, u_2, ..., u_n) \in R_+^n.$$

The entries of the covariance matrix are given by (5.19) and elements of the correlation matrix are like (5.20).

The conditional probability density functions follows from (5.16) and (5.22)

$$
\begin{aligned}
&f(x_{k_1},\dots,x_{k_i} \mid x_{k_{i+1}},\dots,x_{k_n}) \\
&= \{1+[1+\sum_{s=2}^{n-i} \sum_{k_{i+1}\le h_1<\dots<h_s\le k_n} (-1)^s \alpha_{h_1,\dots,h_s} \prod_{j=1}^{s} (1-2\exp(-x_{h_j}^{\beta_{h_j}}))]^{-1} \\
&\times[\sum_{s=2}^{n} \sum_{G} (-1)^s \alpha_{r_1,\dots,r_s} \prod_{j=1}^{s} (1-2\exp(-x_{h_j}^{\beta_{h_j}}))]\} \\
&\times \prod_{j=1}^{i} \beta_j x_j^{\beta_j-1}(1-\exp(-x_j^{\beta_j}))
\end{aligned}
$$

$$G=\{k_1\le r_1<\dots<r_s\le k_n,(r_1,r_2,\dots,r_s)\neq(h_1,h_2,\dots,h_s); s=2,\dots,n -$$

Hence, the regression of X_{k_1} on $X_{k_2},\dots,X_{k_n}$ is

$$
\begin{aligned}
&E(X_{k_1} \mid X_{k_2},\dots,X_{k_n}) \\
&= \{1+[1+\sum_{s=2}^{n-1} \sum_{k_{i+1}\le h_1<\dots<h_s\le k_n} (-1)^s \alpha_{h_1,\dots,h_s} \prod_{j=1}^{s} (1-2\exp(-x_{h_j}^{\beta_{h_j}}))]^{-1} \\
&\times[\sum_{s=2}^{n} \sum_{G} (-1)^s \alpha_{r_1,\dots,r_s} (1-2^{-1/\beta_{k_1}}) \prod_{j=1}^{s} (1-2\exp(-x_{r_j}^{\beta_{r_j}}))]\}\Gamma\left(1+\frac{1}{\beta_{k_1}}\right),
\end{aligned}
$$

where as above

$$G=\{k_1\le r_1<\dots<r_s\le k_n,(r_1,r_2,\dots,r_s)\neq(h_1,h_2,\dots,h_s); s=2,\dots,n-i\}$$

5.7 Further Extensions

In this section, we shall discuss a number of extensions of FGM distributions with uniform marginals designed to increase the maximal value of the correlation coefficient. We thus discuss a number of polynomial copulas.

5.7.1 *Huang and Kotz extensions*

1. In this case, the joint distribution function of bivariate variables (X, Y) is:

$$F_\alpha(x, y) = xy\left[1 + \alpha(1 - x^p)(1 - y^p)\right], \; p > 0, \; 0 \leq x, y \leq 1. \tag{5.23}$$

And the pdf is

$$f_\alpha(x, y) = 1 + \alpha(1 - (1 + p)x^p)(1 - (1 + p)y^p).$$

The admissible range for α in this case is given by

$$-\left(\max\{1, p\}\right)^{-2} \leq \alpha \leq p^{-1}.$$

The range of $corr(X, Y) = 3(\frac{p}{p+2})^2$ is :

$$-3(p + 2)^{-2} \min\left\{1, p^2\right\} \leq \rho \leq \frac{3p}{(p + 2)^2};$$

i.e. for $p = 2$, $\rho_{\max} = \frac{3}{8}$, and for $p = 1$ $\rho_{\min} = -3/16$.

The introduction of parameter p allows us to increase the maximal correlation for the FGM-ditribution with uniform marginals. See Fig. 5.1.

2. Alternative extension Another extension of the bivariate FGM with the uniform marginals is given by

$$F_\alpha(x, y) = xy\left[1 + \alpha(1 - x)^p(1 - y)^p\right], \; p > 0, 0 \leq x, y \leq 1. \tag{5.24}$$

Here the p.d.f. is

$$f_\alpha(x, y) = 1 + \alpha(1 - x)^{p-1}(1 - y)^{p-1}(1 - (1 + p)x)(1 - (1 + p)y)$$

and the admissible range of α is (for $p > 1$):

$$-1 \leq \alpha \leq \left(\frac{p + 1}{p - 1}\right)^{p-1}.$$

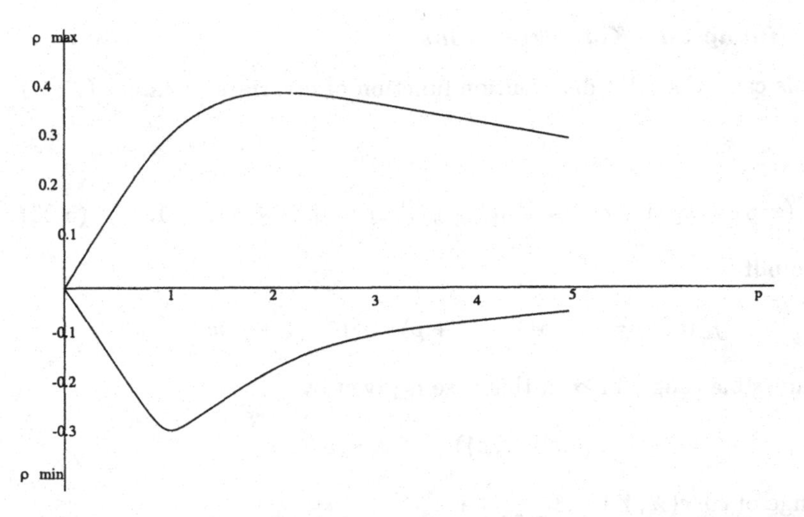

Fig. 5.1 Bounds on correlations ρ as a function of parameter p for the distribution (5.23).

The range is empty for $p < 1$. The $corr(X, Y) = 12\alpha\left(\frac{1}{(p+2)(p+1)}\right)^2$ varies between

$$-12\left(\frac{1}{(p+2)(p+1)}\right)^2 \leq \rho \leq 12\frac{(p-1)^{1-p}(p+1)^{p-3}}{(p+2)^2};$$

i.e. for $p = 1.877$, $\rho_{\max} = 0.3912$, $\rho_{\min} = -0.333$.

Hence, in this case, the maximal positive correlation is even higher than the one attained in the first extension of Huang and Kotz given by Eq. (5.23). See Fig. 5.2.

5.7.2 *Sarmanov's extension*

The density of the joint distribution of the variables X and Y is given here by the expression :

$$\begin{aligned} f(x,y) &= 1 + 3\alpha(2x-1)(2y-1) + \frac{5}{4}\alpha^2\left[3(2x-1)^2 - 1\right]\left[3(2y-1)^2 - 1\right] \\ & \qquad 0 \leq x, y \leq 1 \end{aligned} \tag{5.25}$$

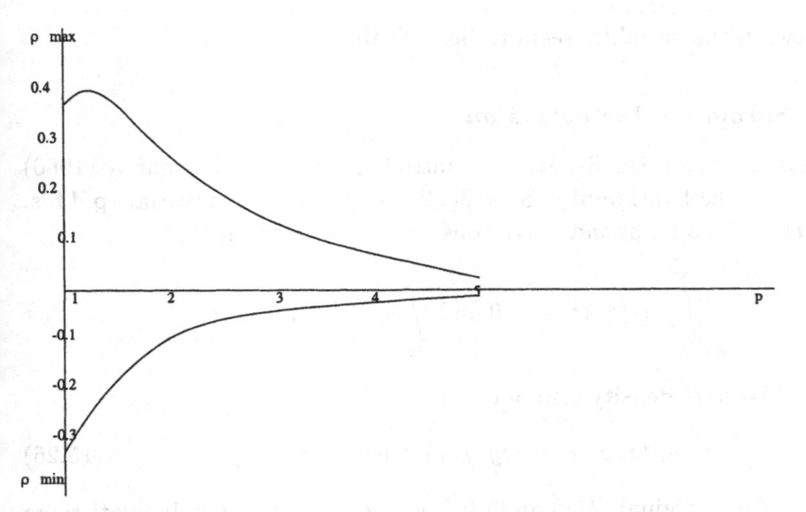

Fig. 5.2 Bounds on correlations ρ as a function of parameter p for the distribution (5.24).

and distribution function is

$$F(x,y) = xy\left\{1+3\alpha(1-x)(1-y)+5\alpha^2(1-x)(1-2x)(1-y)(1-2y)\right\}$$
$$0 \le x, y \le 1.$$

I.O. Sarmanov (1974) makes brief mention of the distribution with density (5.25), obtained using the first two Legendre polynomials. If α satisfies the condition

$$|\alpha| \le \frac{\sqrt{7}}{5} \simeq 0.55$$

then the function (5.25) is nonnegative in square $[0,1]\times[0,1]$ and is a genuine probability distribution. Moreover α is also the coefficient of correlation between X and Y. If we substitute arbitrary distribution functions X and Y, for example, the incomplete gamma functions into (5.25)

$$X = \frac{1}{\Gamma(\alpha+1)}\int_0^{\xi} x^{\alpha}e^{-x}dx,\ Y = \frac{1}{\Gamma(\alpha+1)}\int_0^{\eta} y^{\alpha}e^{-y}dy$$

we shall obtain a new type of two-dimensional gamma correlation surfaces. Further investigations of this family which allows for a rather high correlation between the variables seem to be desirable.

5.7.3 *Sarmanov-Lee extension*

The class of a bivariate distributions introduced by O.V. Sarmanov (1966) encompasses the FGM family. Specifically, let f and g be univariate p.d.f.'s. Let $\psi_i(t), i = 1, 2$ be bounded nonconstant functions such that

$$\int_{-\infty}^{\infty} f(t)\psi_1(t)dt = 0 \text{ and } \int_{-\infty}^{\infty} g(t)\psi_2(t)dt = 0.$$

Define a bivariate density function

$$h_\alpha(x, y) = f(x)g(y)\{1 + \alpha\psi_1(x)\psi_2(y)\} \tag{5.26}$$

with specified marginals $f(x)$ and $g(x)$, where α is a real number satisfying the condition that $1 + \alpha\psi_1(x)\psi_2(y) \geq 0$ for all x and y. The family with joint p.d.f. (5.26) is sometimes called the Sarmanov family of bivariate distributions. This construction is identical with Rüschendorf construction (see Chapter 4 Section 4.5.1), who possibly was not aware of this work, which appeared in a mathematical Russian Journal. Lee (1996) considers "kernels" of type $\psi_1(x) = x - \mu_1, \psi_2(y) = y - \mu_2$, where $\mu_1 = E(X)$, $\mu_2 = E(Y)$, and shows that the range of correlation coefficients for this family of distributions is determined by both the marginal distributions and their mixing functions ψ_i $(i = 1, 2)$ and may therefore be wider than the one for the FGM distributions. With Lee's model we are thus returning to polynomial situation. The range of α in this case is:

$$\max\left(\frac{-1}{\mu_1\mu_2}, \frac{-1}{(1-\mu_1)(1-\mu_2)}\right) \leq \alpha \leq \min\left(\frac{1}{\mu_1(1-\mu_2)}, \frac{1}{\mu_2(1-\mu_1)}\right) \tag{5.27}$$

and correlation coefficient is given by

$$\rho = \frac{\alpha\nu_1\nu_2}{\sigma_X\sigma_Y}, \tag{5.28}$$

where $\nu_1 = \int x\psi_1(x)f(x)dx$, $\nu_2 = \int y\psi_2(y)g(y)dy$ and $\sigma_X = \sqrt{Var(X)}$, $\sigma_Y = \sqrt{Var(Y)}$. However for the *uniform* $[0, 1]$ marginals it follows from (5.27) and (5.28) that

$$-1 \leq \alpha \leq 1$$

and

$$-\frac{1}{3} \leq \rho \leq \frac{1}{3},$$

which is the same range as for the classical FGM. Actually Lee's model leads to the original FGM family.

5.7.4 *Bairamov-Kotz extensions*

Bairamov and Kotz (2000a) investigate a two-parameter extension of the FGM family adding an additional parameter to the Huang-Kotz distribution discussed in Section 5.7.1.

1. The distribution function of (X, Y) is given by :

$$F_{\alpha,a,b}(x,y) = xy\left[1+\alpha(1-x^a)^b(1-y^a)^b\right], a>0, b\geq 1, 0\leq x,y\leq 1,$$

with the pdf

$$f_{\alpha,a,b}(x,y) = 1+\alpha(1-x^a)^{b-1}(1-y^a)^{b-1}\left[1-x^a(1+ab)\right]\left[1-y^a(1+ab)\right].$$

The admissible range of α is: for $b>1$,

$$-\min\left\{1, \left[\frac{1}{a^b}(\frac{ab+1}{b-1})^{b-1}\right]^2\right\} \leq \alpha \leq \left[\frac{1}{a^b}(\frac{ab+1}{b-1})^{b-1}\right],$$

and for $b=1$, the quantity inside the square bracket is to be taken as $\frac{1}{a}$. The maximal and minimal values of $corr(X,Y) = 12\alpha\left[\frac{b}{ab+2}\frac{\Gamma(b)\Gamma(a/2)}{\Gamma(b+\frac{2}{a})}\right]^2$ are within the range

$$-12t^2(b,a)\min\left\{1, \left[\frac{1}{a^b}(\frac{ab+1}{b-1})^{b-1}\right]^2\right\} \leq \rho \leq 12t^2(b,a)\frac{1}{a}\left(\frac{1+ab}{a(b-1)}\right)^{b-1}, \quad (5.29)$$

where $t(x,y) = \frac{\Gamma(x+1)\Gamma(\frac{2}{y})}{y\Gamma(x+1+\frac{2}{y})}$, and $\Gamma(x) = \int\limits_0^\infty t^{x-1}e^{-t}dt-$ is the gamma function.

For $a = 2.8968$ and $b = 1.4908$, we have $\rho_{\max} = 0.5015$.

And for $a = 2$ and $b = 1.5$, $\rho_{\min} = -0.48$.

An alternative extension is a generalization of Huang-Kotz family :

2. Here the distribution function of (X, Y) is

$$F_{p,q,n;\alpha}(x,y) = x^p y^p\left\{1+\alpha(1-x^q)^n(1-y^q)^n\right\}, \quad p,q\geq 1, n>1,$$

$$0 \le x, y \le 1, \quad (5.30)$$

with marginals $F_1(x) = x^p, F_2(y) = y^p, 0 \le x, y \le 1$. The pdf is

$$f_{p,q,n;\alpha}(x,y)$$

$$= x^{p-1}y^{p-1}\left\{p^2 + \alpha(1-x^q)^{n-1}\left[(p - x^q(p+qn)\right](1-y^q)^{n-1} \left[(p - y^q(p+qn))\right]\right\}. \quad (5.31)$$

The admissible range for α:

$$-\min\left\{1, \frac{p^2}{q^2}\left[\frac{p+qn}{q(n-1)}\right]^{2(n-1)}\right\} \le \alpha \le \frac{p}{q}\left[\frac{p+qn}{q(n-1)}\right]^{n-1}$$

and the admissible range for correlation coefficient is

$$-a(n,p,q)\min\left\{1, \frac{p^2}{q^2}\left[\frac{p+qn}{q(n-1)}\right]^{2(n-1)}\right\} \le \rho \le a(n,p,q)\frac{p}{q}\left[\frac{p+qn}{q(n-1)}\right]^{n-1},$$

where

$$a(n,p,q) = \left(\frac{p}{(p+2)(p+1)^2}\right)^{-1}\frac{1}{q^2}\left(Beta(\frac{p+1}{q}, n+1)\right)^2.$$

For $p = 1$, $q = 2$ the maximal negative correlation $\rho_{\min} = -0.4794$ is attained at $n = 1.495$. For $p = 0.001$, $q = 1.5$ the maximal positive correlation $\rho_{\max} = 0.6122$ is attained at $n = 1.379$. It should however be noted that (5.30) is not a copula.

5.7.5 *Lai and Xie extension*

Lai and Xie (2000) consider a bivariate function

$$C(u,v) = uv + w(u,v) = uv + \alpha u^b v^b (1-u)^a (1-v)^a, \quad a, b \ge 1, \quad (5.32)$$

(which is, indeed a particular case of the Rüschendorf construction) and have shown that (5.32) is PQD for $0 \le \alpha \le 1$.

Bairamov and Kotz (2000b) observe that (5.32) is a bivariate distribution function for α over a wider range satisfying

$$-\min\left\{\frac{1}{[B^+(a,b)]^2}, \frac{1}{[B^-(a,b)]^2}\right\} \le \alpha \le -\frac{1}{B^+(a,b)B^-(a,b)},$$

and possessing the PQD property for α satisfying

$$0 \leq \alpha \leq -\frac{1}{B^+(a,b)B^-(a,b)},$$

where

$$\begin{aligned} B^+(a,b) &= \left[\frac{b(a+b-1)+\sqrt{ab(a+b-1)}}{(a+b)(a+b-1)}\right]^{b-1} \\ &\times \left[1-\frac{b(a+b-1)+\sqrt{ab(a+b-1)}}{(a+b)(a+b-1)}\right]^{a-1} \\ &\times \left[(a+b)\frac{b(a+b-1)+\sqrt{ab(a+b-1)}}{(a+b)(a+b-1)}-b\right] \end{aligned} \quad (5.33)$$

and

$$\begin{aligned} B^-(a,b) &= \left[\frac{b(a+b-1)-\sqrt{ab(a+b-1)}}{(a+b)(a+b-1)}\right]^{b-1} \\ &\times \left[1-\frac{b(a+b-1)-\sqrt{ab(a+b-1)}}{(a+b)(a+b-1)}\right]^{a-1} \\ &\times \left[(a+b)\frac{b(a+b-1)-\sqrt{ab(a+b-1)}}{(a+b)(a+b-1)}-b\right]. \end{aligned} \quad (5.34)$$

For example letting $a = 5$ and $b = 5$ we have from (5.33) and (5.34) $-\frac{1}{B^+(a,b)B^-(a,b)} = 6.053 \times 10^4$ which is by far larger than 1. Further investigations of the Lai and Xie family could be fruitful.

5.7.6 *Bairamov-Kotz-Bekçi generalization*

Our final generalization of the FGM family involves four parameters. The distribution function of (X, Y) is

$$\begin{aligned} F(x,y) &= xy\left\{1+\alpha\left(1-x^{p_1}\right)^{q_1}\left(1-y^{p_2}\right)^{q_2}\right\} \quad p_1, p_2 \geq 1\,,\ q_1, q_2 \geq 1 \\ & \quad 0 \leq x, y \leq 1. \end{aligned} \quad (5.35)$$

The pdf is

$$f(x,y) = 1 + \alpha (1 - x^{p_1})^{q_1 - 1} [1 - (1 + p_1 q_1) x^{p_1}]$$

$$\times (1 - y^{p_2})^{q_2 - 1} [1 - (1 + p_2 q_2) y^{p_2}] \quad 0 \leq x, y \leq 1. \tag{5.36}$$

The admissible range of α is

$$-\min \left\{ 1, \frac{1}{p_1 p_2} \left(\frac{1 + p_1 q_1}{p_1 (q_1 - 1)} \right)^{q_1 - 1} \left(\frac{1 + p_2 q_2}{p_2 (q_2 - 1)} \right)^{q_2 - 1} \right\} \leq \alpha$$

$$\leq \min \left\{ \frac{1}{p_1} \left(\frac{1 + p_1 q_1}{p_1 (q_1 - 1)} \right)^{q_1 - 1}, \frac{1}{p_2} \left(\frac{1 + p_2 q_2}{p_2 (q_2 - 1)} \right)^{q_2 - 1} \right\}.$$

The admissible range of $corr(X,Y) = \rho = \frac{Cov(X,Y)}{\sigma_X \sigma_Y} = 12\alpha t(q_1, p_1) t(q_2, p_2)$ is

$$-12t(q_1, p_1) t(q_2, p_2) \min \left\{ 1, \frac{1}{p_1 p_2} \left(\frac{1 + p_1 q_1}{p_1 (q_1 - 1)} \right)^{q_1 - 1} \left(\frac{1 + p_2 q_2}{p_2 (q_2 - 1)} \right)^{q_2 - 1} \right\}$$

$$\leq \rho \leq$$

$$12t(q_1, p_1) t(q_2, p_2) \min \left\{ \frac{1}{p_1} \left(\frac{1 + p_1 q_1}{p_1 (q_1 - 1)} \right)^{q_1 - 1}, \frac{1}{p_2} \left(\frac{1 + p_2 q_2}{p_2 (q_2 - 1)} \right)^{q_2 - 1} \right\},$$

where $t(.,.)$ is defined in section 5.7.4 above Eq. (5.29). Clearly (5.35) is the most general form of modified FGM distributions.

It is of interest whether (5.35) satisfies TP_2 property for all $p_1, p_2 \geq 1$, $q_1, q_2 \geq 1$ or not. The answer is negative and contained in the following example.

The condition allowing to be TP_2 is that for all $x_1 < x_2$ and $y_1 < y_2$

$$f(x_1, y_1) f(x_2, y_2) - f(x_1, y_2) f(x_2, y_1) \geq 0. \tag{5.37}$$

For the distribution (5.36) denoting $a(x; p, q) = [1 - x^p]^{q-1} [1 - (1 + pq) x^p]$ (5.37) can be written as follows:

$$x_1 y_1 \{1 + \alpha a(x_1; p_1, q_1) a(y_1; p_2, q_2)\} x_2 y_2 \{1 + \alpha a(x_2; p_1, q_1) a(y_2; p_2, q_2)\}$$

$$-x_1y_2\{1+\alpha a(x_1;p_1,q_1)a(y_2;p_2,q_2)\}\cdot x_1y_1\{1+\alpha a(x_2;p_1,q_1)a(y_1;p_2,q_2)\}\geq 0$$

$$x_1y_1y_1y_2\alpha[a(x_1;p_1,q_1)a(y_1;p_2,q_2)+a(x_2;p_1,q_1)a(y_2;p_2,q_2)$$

$$-a(x_1;p_1,q_1)a(y_2;p_2,q_2)-a(x_2;p_1,q_1)a(y_1;p_2,q_2)]\geq 0$$

$$x_1y_1y_1y_2\alpha\{a(x_1;p_1,q_1)[a(y_1;p_2,q_2)-a(y_2;p_2,q_2)]$$

$$-a(x_2;p_1,q_1)[a(y_1;p_2,q_2)-a(y_2;p_2,q_2)]\}\geq 0$$

or

$$x_1y_1y_1y_2\alpha[a(x_1;p_1,q_1)-a(x_2;p_1,q_1)]$$

$$\times[a(y_1;p_2,q_2)-a(y_2;p_2,q_2)]\geq 0. \tag{5.38}$$

But if $f(x)=1$, $0\leq x\leq 1$, i.e. if the marginals are uniform distributed on $[0,1]$, for $\alpha=1$, $p_1=2$, $p_2=2$, $q_1=2$, $q_2=2$ and $x_1=0.1$, $x_2=0.2$, $y_1=0.7$, $x_1=0.9$ from (5.38) one has

$$f(x_1,y_1)f(x_2,y_2)-f(x_1,y_2)f(x_2,y_1)=-0.0276\leq 0,$$

i.e. X and Y are not TP_2, whenever classical FGM is TP_2 by the Lemma 1.1. See Bairamov *et al.* (2000) for further details.

5.7.7 *Concomitants of order statistics*

Let (X_i,Y_i), $i=1,2,...,n$ be a random sample from an absolutely continuous bivariate population (X,Y) with a d.f. $F_{X,Y}(x,y)$. Let $X_{r:n}$ denote the rth order statistics of the X sample values. Denote by $Y_{[r:n]}$ the Y value associated with $X_{r:n}$. We call $Y_{[r:n]}$ the concomitant of the rth order statistic. For more details the reader is referred to the review articles of Bhattacharya (1984) and David (1993). More recently Balasubramanian and Beg (1997), (1998) have studied concomitants in FGM family with exponential marginals. Denote probability density function (p.d.f.) of $Y_{[r:n]}$ by $g_{[r:n]}(y)$. It is known that

$$g_{[r:n]}(y)=\int_{-\infty}^{+\infty}f(y\mid x)\,f_{r:n}(x)\,dx$$

where $f(y \mid x)$ is the conditional density function of Y, given X and $f_{r:n}(x)$ is the p.d.f. of $X_{r:n}$ (see David (1981)). The most important use of concomitants arises in selection procedures when r $(1 \leq r \leq n)$ individuals are chosen on the basis of their X values. Then the corresponding Y values represents performance on an associated characteristic.

The distribution and recurrence relation between moments of concomitants in the classical bivariate FGM distribution with uniform marginals are given below. The d.f. and p.d.f. of $Y_{[r:n]}$ are respectively

$$G_{[r:n]}(y) = y\left\{1 + \alpha\left[1 - 2\frac{r}{n+1}\right](1-y)\right\}, 0 \leq y \leq 1 \tag{5.39}$$

and

$$g_{[r:n]}(y) = 1 + \alpha\left[1 - 2\frac{r}{n+1}\right](1-2y). \tag{5.40}$$

See Bairamov and Bekçi (1999) for additional details.

For the generalized distribution (5.35), we have the following result :

Theorem. a) The kth moment of $Y_{[r:n]}$ is

$$\mu_{[r:n]}^{(k)} = \frac{1}{k+1} + \alpha C(k; p_2, q_2) K(r, n, p_1, q_1).$$

Thus the expected value and variance of $Y_{[r:n]}$ are

$$E\left\{Y_{[r:n]}\right\} = \frac{1}{2} + \alpha C(1; p_2, q_2) K(r, n, p_1, q_1)$$

and

$$Var\left\{Y_{[r:n]}\right\} = \frac{1}{12} + \alpha K(r, n, p_1, q_1)$$

$$\times \left\{C(2; p_2, q_2) - C(1; p_2, q_2)\left[1 + \alpha C(1; p_2, q_2) K(r, n, p_1, q_1)\right]\right\},$$

respectively. The moment generating function of $Y_{[r:n]}$ is

$$M_{[r:n]}(t) = E\left\{e^{tY_{[r:n]}}\right\} = \int_{-\infty}^{+\infty} e^{ty} g_{[r:n]}(y)\, dy$$

$$= \frac{e^t - 1}{t} + \alpha K(r, n, p_1, q_1) \sum_{j=0}^{q_2-1} (-1)^j \binom{q_2 - 1}{j} \left[S(jp_2) - (1 + p_2 q_2) S((j+1)p_2\right.$$

Here $B(a,b)$ is the beta function : $B(a,b) = \int_0^1 t^{a-1}(1-t)^{b-1}dt,\ a>0,\ b>0$ and

$$K(r,n,p,q) = \sum_{i=0}^{q-1}(-1)^i \binom{q-1}{i} \frac{1}{B(r,n-r+1)}$$

$$\times[B(ip+r,n-r+1)-(1+pq)B((i+1)p+r,n-r+1)]\ ,\ (1 \le r \le n, p,\ 1 \le q)$$

$$C(k;p,q) = \tfrac{1}{p}\left[B\left(\tfrac{k+1}{p},q\right) - (1+pq)B\left(\tfrac{k+1}{p},q\right)\right],$$
$$(k = 0,1,2,...;\ 1 \le p,\ q)$$

$$S(a) = \frac{e^t}{t}\left[\Sigma_{k=0}^{a}\frac{(-1)^k a!}{t^k(a-k)!}\right] + (-1)^{a+1}\frac{a!}{t^{a+1}}\ .$$

5.8 FGM Sequences

(1) Limiting distribution
Define a FGM distribution F in R^n, for $n \ge 1$ with respect to the given univariate distributions F_i, $i \le n$, by

$$F(x_1,x_2,...,x_n) = \prod_{i=1}^{n} F_i(x_i)\left\{1 + \sum_{1<j<k\le n} \alpha(j,k)S_j(x_j)S_k(x_k)\right\} \tag{5.41}$$

for all vectors $\mathbf{x} = (x_1,x_2,...,x_n) \in R^n$, where the $n(n-1)/2$ terms $\alpha(j,k)$ are suitable constants, such that F is a distribution function, where $S_j(x_j) = 1 - F_j(x_j)$. This is a particular case of a general FGM distribution, discussed in Section 5.6, where $\alpha(j,k,l) = 0$, $\alpha(j,k,l,m) = 0$ and so on. The univariate marginals of F are the F_i. The constants $a(j,k)$ are admissible if the 2^n inequalities

$$1 + \sum_{1\le j<k\le n} \alpha(j,k)\xi_j\xi_k \ge 0 \tag{5.42}$$

hold, for all $\xi_j = -M_j$ or $1 - m_j$, where

$$M_j = \sup\{F_j(x), -\infty < x < \infty\} \setminus [0,1]$$

and $m_j = \inf\{F_j(x), -\infty < x < \infty\} \setminus [0,1]$. If F_j is absolutely continuous, then $M_j = 1$ and $m_j = 0$, hence $\xi_j = \pm 1$.
Hashorva and Hüsler (1999) define a FGM random sequence $\{X_j, j \geq$ with univariate marginals $X_i \sim F_i$, $i \geq 1$, and a symmetric function $\alpha(.,.)$ $(\alpha(j,k) = \alpha(k,j))$ such that the joint distribution of $X_{i_1}, X_{i_2}, ..., X_{i_n}$ is given by the FGM distribution

$$F_{i_1,i_2,\ldots,i_n}(\mathbf{x}) = \prod_{i=1}^{n} F_{i_k}(x_i) \left\{ 1 + \sum_{1 \leq j < k \leq n} \alpha(i_j, i_k) S_{i_j}(x_j) S_{i_k}(x_k) \right\}. \tag{5.43}$$

The parameters $\alpha(.,.)$ are admissible if for every $n \geq 1$ and $\{i_1, i_2, ...,$ the inequalities

$$1 + \sum_{1 \leq j < k \leq n} \alpha(i_j, i_k) \xi_{i_j} \xi_{i_k} \geq 0$$

hold for all ξ_{i_j}. The FGM sequence is stationary iff the univariate marginals are all equal,

$$F_i = F_1, \; i > 1,$$

and the parameters $\alpha(j,k)$ depend on j,k only through their difference, i.e.:

$$\alpha(j,k) = \alpha(j-k) \text{ for all } \; j \neq k.$$

Let also $\lim_{n\to\infty} \sup_{j-k>n} |\alpha(j,k)| \to 0$ Consider partial maxima $M_n = \max_{i \leq n} X_i$, $n \geq 1$. Let $\{X_i, i \geq 1\}$ be a sequence of random variables X_i whose joint finite dimensional distributions are FGM. Hashorva and Hüsler (1999) derived the limiting distribution of the maxima with respect to some suitable normalization $u_n(x)$. They consider an approximation to $P\{M_n \leq u_n(x)\}$. In the case under consideration the dependence between the random variables X_i is not very strong and the approximation

$$P\{M_n \leq u_n(x)\} \approx \prod_{i \leq n} P\{X_i \leq u_n(x)\} = \prod_{i \leq n} F_i(u_n(x))$$

is valid. Suppose that the normalization $u_n(x)$ satisfies

$$\lim_{n\to\infty} \sup_{i \leq n} F_i(u_n(x)) = 0 \tag{5.44}$$

for the set of x with $\liminf_{n\to\infty}\prod_{i\le n}F_i(u_n(x))>0$. This condition is essential for general results in the general case of nonidentically distributed X_i. More precisely, suppose that $\{X_i, i\ge 1\}$ be an FGM sequence such that (5.44) as well as

$$\sup_{j-k>n}|\alpha(j,k)|\to 0, n\to\infty$$

and

$$\limsup_{n\to\infty}\sum_{i\le n}S_i(u_n(x))<\infty$$

hold with respect to a normalization $u_n(x)$. Then

$$P\{M_n\le u_n(x)\}-\prod_{i\le n}F_i(u_n(x))\to 0$$

as $n\to\infty$. In addition if

$$\lim_{n\to\infty}\prod_{i\le n}F_i(u_n(x))=G(x)$$

then

$$P\{M_n\le u_n(x)\}\xrightarrow{d}G(x)$$

as $n\to\infty$. Compare with Joe [111], page 178, who studied related limiting distributions in the context of a distribution of extreme value theory.

(2) Order statistics
Let $X_1, X_2, ..., X_n, ...$ be a sequence of uniform FGM random variables such that for any $n>1$

$$F_n(x_1,x_2,...,x_n)=\prod_{i=1}^{n}x_i\left\{1+\sum_{1\le j<k\le n}\alpha_n(j,k)(1-x_j)(1-x_k)\right\}, \tag{5.45}$$

where $n(n-1)/2$ terms $\alpha_n(j,k)$ are suitable constants. Since F_n is a distribution function,

$$1+\sum_{1\le j<k\le n}\alpha_n(j,k)\xi_j\xi_k\ge 0 \tag{5.46}$$

holds, where $\xi_j = 1$ or -1. Let $\alpha_n(j,k) = \alpha_n$ for all j,k. Then for any n the random variables $X_1, X_2, ..., X_n$ are exchangeable. Namely

$$(X_{i_1}, X_{i_2}, ..., X_{i_n}) \stackrel{d}{=} (X_{j_1}, X_{j_2}, ..., X_{j_n}).$$

It follows from (5.46) that in this case the admissible range for α_n, allowing (5.45) to be a n–variate copula, is

$$-\frac{1}{\binom{n}{2}} \le \alpha_n \le \frac{1}{\left[\frac{n}{2}\right]},$$

where $[a]$ denotes the integer part of the number a. In this case the coefficients α_n, allowing (5.45) to be a copula converges to 0, for $n \to \infty$, i.e. for large n the FGM sequence becomes independent. (One can call asymptotically independent.)
Exchangeable FGM random variables were originally studied by Shaked (1975) [200]. Consider now the finite FGM sequence $X_1, X_2, ..., X_n$. Denote by $X_{1:n} \le X_{2:n} \le ... \le X_{n:n}$ the order statistics of $X_1, X_2, ..., X_n$. It is known that for symmetrically dependent random variables

$$P\{X_{r:n} \le x\} = \sum_{l=r}^{n} (-1)^{l-r} \binom{l-1}{r-1} \binom{n}{l} P\{X_{l:l} \le x\}.$$

For (5.45) we have

$$P(X_{r:n} \le x) = F_{r:n}(x) = \textstyle\sum_{l=r}^{n} (-1)^{l-r} \binom{l-1}{r-1} \binom{n}{l} x^l \left\{1 + \alpha_n \frac{l(l-1)}{2} (1-x)^2\right\}, \tag{5.47}$$

since for any integer m,

$$P\{X_{m:m} \le x\} = x^m \left\{1 + \alpha_m \frac{m(m-1)}{2} (1-x)^2\right\}.$$

Let $X_{n+1}, X_{n+2}, ..., X_{n+m}$ be a new FGM sample fitting to the same model which is assumed to be independent of $X_1, X_2, ..., X_n$. Then

$$P\{X_{n+1} < X_{r:n}\} = 1 - P\{X_{n+1} > X_{r:n}\}$$

$$= 1 - \int_0^1 P\{X_{n+1} > X_{r:n} \mid X_{n+1} = x\}\, dx = 1 - \int_0^1 P\{X_{r:n} \le x\}\, dx$$

$$= 1 - \sum_{m=r}^{n} (-1)^{m-r} \binom{m-1}{r-1} \binom{n}{m} \left\{ \frac{1}{m+1} + \alpha_n \frac{m(m-1)}{(m+1)(m+2)(m+3)} \right\}.$$

Denote

$$\xi_i = \begin{cases} 1 \ , & X_{n+i} < X_{r:n} \\ 0 \ , & X_{n+i} \ge X_{r:n} \end{cases} \qquad \nu_m = \sum_{i=1}^{m} \xi_i \ .$$

It is evident that ν_m is the number of $X_{n+1}, X_{n+2}, ..., X_{n+m}$ falling below the random threshold $X_{r:n}$.

Bairamov and Eryılmaz (2000) derived the exact distribution of ν_m. Their result is as follows. For any integer $m \ge 1$ and $1 \le r \le n$

$$
\begin{aligned}
&P\{\nu_m = k\} \\
&= \tbinom{m}{k} \sum_{s=r}^{n} (-1)^{s-r} \tbinom{s-1}{r-1} \tbinom{n}{s} \Big[sB(s+k, m-k+1) \\
&+\alpha_n \left(\tfrac{s^2(s-1)}{2} B(s+k, m-k+3) - s(s-1)B(s+k+1, m-k+2) \right) \\
&+\alpha_m \left(\tfrac{sk(k-1)}{2} B(s+k, m-k+3) - sk(m-k)B(s+k+1, m-k+2) \right. \\
&\left. + \tfrac{s(m-k)(m-k-1)}{2} B(s+k+2, m-k+1) \right) \\
&+\alpha_n\alpha_m \left(\tfrac{s^2(s-1)}{2} \tfrac{k(k-1)}{2} B(s+k, m-k+5) - \tfrac{s^2(s-1)}{2} k(m-k) \right. \\
&\times B(s+k+1, m-k+4) + \tfrac{s^2(s-1)}{2} \tfrac{(m-k)(m-k-1)}{2} \\
&\times B(s+k+2, m-k+3) \\
&-s(s-1)\tfrac{k(k-1)}{2} B(s+k+1, m-k+4) + k(m-k)s(s-1) \\
&\times B(s+k+2, m-k+3) \\
&\left. -\ s(s-1)\tfrac{(m-k)(m-k-1)}{2} B(s+k+3, m-k+2) \right) \Big],
\end{aligned}
$$

$$k = 0, 1, ..., m \quad , \quad -\frac{1}{\binom{m}{2}} \le \alpha_m \le \frac{1}{\left[\frac{m}{2}\right]}. \tag{5.48}$$

Here, as above, $B(a, b)$ is the Beta function. Although expression 5.48 looks formidable, calculations are actually quite straightforward. This distribution is of importance in many practical problems when we are interested in the behavior of sequence of observations in relation to a certain order statistic (minimum, median, quantile, maximum) from a given population.

Below we present some numerical values of $P\{\nu_m = k\}$ for selected values of the parameters n, r, m, α_n and α_m. These results are quite informative. (Compare, in particular, the first and third rows.)

Table 1. Numerical values of $P\{\nu_m = k\}$.

n	r	m	α_n	α_m	k	$P\{\nu_m = k\}$
3	1	2	0.5	−0.25	0	0.629
					1	0.267
					2	0.104
4	2	2	−0.1	−0.5	0	0.403
					1	0.422
					2	0.175
5	2	3	0.1	0.75	0	0.347
					1	0.310
					2	0.231
					3	0.112

Chapter 6

Global Versus Local Dependence between Random Variables

6.1 Introduction

Global indices were used and continue to be used in probability theory and statistics for many years to measure the dependence between two random variables X and Y. As it was emphasized in Chapter 2 among these indices the linear correlation coefficient, Spearman's correlation coefficient on ranks, Kendall's concordance coefficient are by far the most prominent. The linear correlation coefficient is appropriate mainly for normal variables, the two others are independent of the marginal distributions of X and Y. Some authors have measured the dependence between two random variables by mutual information or relative entropy. In the case of normal variables relative entropy is connected with the absolute value of the linear correlation coefficient.

These indices are defined from the moments of the distribution on the whole plane and can be zero when X and Y are not independent. One needs therefore indices which measure the dependence locally. In the case where X and Y are survival variables, one needs to identify the time of maximal association : for example the delay before the first symptom of a genetic disease by members of the same family will appear. The pairs (X, Y) and (X', Y') can have the same global measure of dependence but may possess two different distributions F and F' : a local index will allow us to compare them, and in the case of survival variables to compare their variations in time. The variations with x and y of some local indices allow us to characterize certain distributions and conversely choosing a shape of variation for an index allows us sometimes to choose an appropriate model.

In the second Section we reproduce several axioms established by Renyi [179] that such a measure should satisfy. We then give several definitions of three global indexes, and their connection with concepts of dependence. This has also been studied in the papers of Genest [81], Long and Krzysztofowicz [147], and Nelsen [160]. We next provide several definitions of mutual information or relative entropy. These notions have been established by Bell [20], Joe [108], and illustrated by Kapur [121] for non-normal distributions and multivariate vectors. Starting from a characteristic property of the relative entropy for a $p+q$ random normal vectors (X, Y), Lin [145], and later Zografos [233] have developed new measures of dependence constructed with the canonical correlation coefficient of the two vectors X and Y. The third Section defines a few local indices studied by Bjerve, Doksum and Blyth [21], [61], [25], Dabrowska [53], Clayton [46] and Oakes [163], Holland and Wang [98] and Jones [118] and their connections with certain concepts of dependence. The fourth Section explains how to obtain a non-parametric estimation of two local indices, and the properties of these estimations. Dabrowska [53] has developed these notions. We are searching for localization of maximal dependence in the fifth section. This is illustrated by the distributions of three copulas with the same concordance coefficient, and the evolutions of three local indices for these distributions. A part of this work was originally presented in Drouet [62]. A number of results scattered in the literature are unified.

6.2 Global Measures of Dependence

6.2.1 *Desirable properties of a measure of dependence*

Renyi [179], [180] and Bell[20] have discussed some axioms that global measures should satisfy; these properties are also desirable for local measures. Note that among the indices presented here, some are not measures, since they may be negative. The desirable properties should be :

(1) Standardization
The values of an index are between 0 and 1.
(2) Independence
If X and Y are independent, the index should be zero.
(3) Functional dependence
If one variable is a function of the other, the index must be equal

to 1.

(4) Increasing property
The index should increase as the dependence increases.

(5) Invariance
The index should be invariant with respect to a linear transformation of the variables. A stronger condition would be that the index is marginally free, i.e. the measure of the dependence is the same as the corresponding measure on the copula.

(6) Symmetry
If the variables are exchangeable, then the index should be symmetric.

(7) Relationship with measures for ordinal variables.
If the index is defined for both ordinal and continuous variables, there should be a close relationship between the two measures.

(8) Interpretability. This is a very delicate and intangible property. Roughly speaking, it means that the numerical value of this index can be translated into a qualitative meaningful measure.

6.2.2 *Covariance, Q-covariance*

To measure the dependence between two random variables X and Y one may use the covariance:

$$cov(X,Y) = E(XY) - E(X).E(Y)$$

provided that X, Y and $X.Y$ are integrable functions. Hoeffding (1940) has extended this concept to a larger class of random variables by the formula:

$$cov(X,Y) = \int_{-\infty}^{+\infty}\int_{-\infty}^{+\infty}(F(x,y) - F_1(x).F_2(y))dxdy$$

valid when the above formula is finite. Krajka and Szynal [136] have shown that the concept of covariance may be further extended using the property of quantiles when one variable, for example Y is not integrable. Suppose that X is integrable. Let $Y(p)$ be the quantile function of Y, and I be the indicator function. For X integrable and X and Y with continuous distributions functions, write :

$$L_{X,Y}(p) = E\left((X - E(X))I(Y \geq y(p))\right).$$

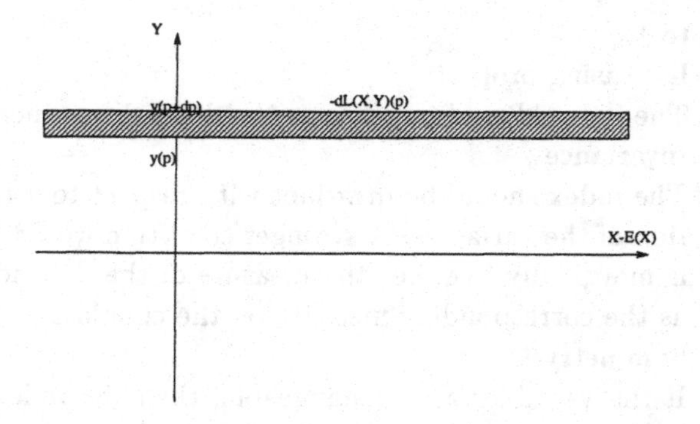

Fig. 6.1 The function $L_{X,Y}(p)$.

$$\overline{L}_{X,Y}(p) = E(X - E(X)).I(Y < y(p)).$$

Similarly one defines $L_{Y,X}(p)$ and $\overline{L}_{Y,X}(p)$ for Y integrable and not necessarily integrable X.

The Q-covariance is defined by the change of variables $(X,Y) \mapsto (X,p)$, where p is the quantile associated with Y :

$$cov^Q(x,y) = -\int_0^1 y(p)dL_{X,Y}(p) = -\int_0^1 y(p)d\overline{L}_{X,Y}(p),$$

whenever one integral is finite. A more interpretable formula is given by :

$$cov^Q(x,y) = E\left(Xy(P)(\frac{f(X,y(P)}{f_1(x)f_2(y(P)} - 1)\right)$$

whenever $Xy(P)(\frac{f(X,y(P)}{f_1(x)f_2(y(P)} - 1)$ is integrable. Here P is a uniform variable on $[0,1]$ independent of X, and $f(.,.)$ is the density of the pairs (X,Y). When the two variables X and Y are integrable the three definitions coincide :

$$cov(X,Y) = cov^Q(X,Y) = cov^Q(Y,X).$$

Examples : Pairs of random variables (X,Y), where Y has a Cauchy density function

$$f(y) = \frac{1}{\pi(1+y^2)}, \ -\infty < y < +\infty$$

or also the density $f(y) = \frac{r\alpha^r}{(y+\alpha)^{r+1}}$, $y \geq 0$, $r \geq 1$, which has no moments up to order $r-1$. For this latter distribution, the quantiles are easily obtained. Namely $y(p) = \frac{\alpha p}{1-p}$. Note that when $r = 1$, this distribution is obtained as the ratio of two exponential variables (Lachenbruch and Brogan (1971) [137]).

6.2.3 *The coefficient of linear correlation ρ*

Although we have discussed this coefficient in Chapter 2, at the risk of being redundant we shall briefly review this concept. This is needed to present further promising generalization. The coefficient ρ is formally defined as:

$$\rho = corr(X,Y) = \frac{cov(X,Y)}{\sqrt{var(X)var(Y)}}.$$

ρ is normalized, that is:

$$-1 \leq \rho \leq 1.$$

The two limits (-1) and $+1$ are attained in case when the dependence is maximal (negative or positive respectively).

6.2.3.1 *The case when (X,Y) is bivariate normal*

If the distribution of (X,Y) is a standardized bivariate normal (with mean 0 and variance equal to one for both X and Y), then the conditional density of Y given x is:

$$f(y/X = x) = \frac{1}{\sqrt{2\pi(1-\rho^2)}} \exp\left(-\frac{(y-\rho x)^2}{2(1-\rho^2)}\right).$$

Therefore:

$$E(Y/X = x) = \rho x$$

and

$$var(Y/X = x) = 1 - \rho^2.$$

In this case ρ characterizes the distribution.

We know also that in the family of the bivariate normal distribution, the condition $\rho = 0$ is equivalent to (X,Y) being independent. The equivalence is not valid if the distribution is not normal (except for the case of binary variables X and Y).

For other distributions, the connection between $\mu(x) = E(Y/X = x)$ and x may not be linear and the variance $\sigma^2(x) = var(Y/X = x)$ may not be constant. Hence ρ is not necessarily an adequate index to synthesize the relationship between X and Y. To characterize these connections for non-normal distributions, Bjerve and Doksum [21] [61] and Blyth [25] (cf Section 6.3.4) have defined a local linear correlation coefficient using the functions $\beta(x) = \frac{\partial \mu(x)}{\partial x}$ and $\sigma^2(x)$.

For certain distributions the index ρ may not even exist. For example, consider the bivariate Pareto distribution, described by Mardia [150] with the density function:

$$f(x,y) = \frac{a(a+1)}{\theta_1 \theta_2} . (\frac{x}{\theta_1} + \frac{y}{\theta_2} - 1)^{-a+2} \tag{6.1}$$

where $x \geq \theta_1 > 0$ and $y \geq \theta_2 > 0$ and $a > 0$. The marginal density function for X is:

$$f_1(x) = a.\theta_1^a x^{-a+1}, \; x \geq \theta_1 > 0, \; a > 0. \tag{6.2}$$

It is easy to verify that :

$$E(X) = a.\theta_1 \frac{1}{a-1}, \; a > 1$$

$$E(X^2) = a.\theta_1^2 \frac{1}{a-2}, \; a > 2$$

$$var(X) = a.\theta_1^2 \frac{1}{(a-2)(a-1)^2}, \; a > 2$$

$$cov(X,Y) = \theta_1 \theta_2 \frac{1}{(a-2)(a-1)^2}, \; a > 2$$

therefore $\rho = \frac{1}{a}$, $a > 2$ and does not exist when $0 < a \leq 2$. The maximum correlation coefficient to be discussed next does not have this drawback.

6.2.3.2 *Correlation and extremal properties of normal distributions*

The result proved by Klaassen and Wellner [131] states that for any bivariate normal distribution (X, Y), the maximum correlation coefficient, taken over all the transformations a and b from R to R :

$$\rho_M(X, Y) = sup_{a,b}\rho(a(X), b(Y))$$

is equal to the absolute value of the linear correlation coefficient $|\rho|$.

If a and b are restricted to monotone functions, Kimeldorf and Sampson [130] have proved that the maximum monotone correlation coefficient ρ_m is again :

$$\rho_m(X, Y) = |\rho|.$$

In fact Klaassen and Wellner show that the maximum ρ_M is attained if and only if a and b are linear transformations of X and Y.

This result extends straightforward to the normal copula model, that is the bivariate distributions (X, Y) with any marginal distributions $G(x)$ and $H(y)$ and such that $(\Phi^{-1}(G(X)), \Phi^{-1}(H(Y)))$ has the standardized bivariate normal distribution with correlation coefficient ρ :

$$\rho_M(X, Y) = \rho_M\left(\Phi^{-1}(G(X)), \Phi^{-1}(H(Y))\right) = |\rho|.$$

6.2.3.3 *ρ and the moment of inertia around the line $D_1 : \{y = x\}$*

In the family of the distributions which are bivariate normal with mean zero and variance equal to one, the upper Fréchet bound $F^+(x, y)$ is the line $D_1 : \{y = x\}$. Indeed, in that case $\rho = 1$, therefore $E(Y/X = x) = x$, and $var(Y/X = x) = 0$, hence the distribution of $F^+(x, y)$ is concentrated on the line $D_1 : \{y = x\}$.

If (X, Y) belongs to this family with correlation ρ, then $(1 - \rho)$ is the expectation of the moment of inertia about the line $\{y = x\}$. Indeed, let (X, Y) be an arbitrary point on the plane then the moment of inertia d^2

$$d^2[(X, Y), D_1] = (X - Y)^2/2.$$

Therefore:

$$E(d^2) = 2(E(X^2) - 2E(XY) + E(Y^2))/2 = 1 - \rho.$$

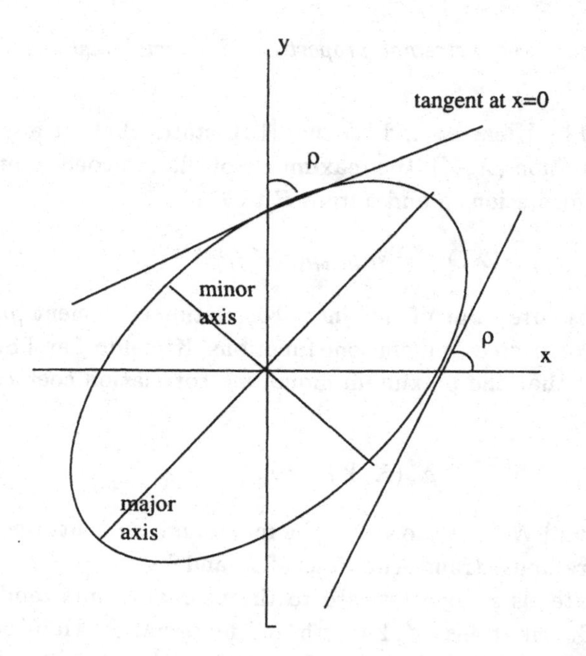

Fig. 6.2 ρ is the slope of the tangent to the ellipse at $x = 0$.

6.2.3.4 *A geometric interpretation*

Among numerous geometric representations of the correlation coefficients which actually go as far as Galton (1885), we shall cite the following recent one: consider two standardized variables (X^*, Y^*) ; Chatillon (1984) provides a rather wide class of bivariate distributions that have elliptically equi-density contours. Each ellipse is centered at the origin and the main axis of the ellipse coincides with the main diagonal of the plane of the domain of definition of these variables if $\rho > 0$, and the other diagonal when $\rho < 0$. Geometrically coefficient ρ is the slope of the tangent to the ellipse at the point $x = 0$. Note that due to the symmetry of the correlation coefficient, we can retrieve the same slope by appropriately interchanging X and Y indicated in Figure 6.2 which extends Figure 3 of Rodgers and Nickwander (1988).

6.2.3.5 *ρ and concepts of dependence*

If the distribution of (X, Y) satisfies any of the concepts of positive dependence, for example if only the weak concept of Positive Quadrant Dependence (PQD) (cf Chapter 3) is imposed, then ρ will nevertheless be positive. Indeed in that case $cov(X, Y) \geq 0$ (Hoeffding lemma). If ρ is positive and (X, Y) is bivariate normal, then (X, Y) satisfies a stronger condition of Likelihood Ratio Dependence (LRD) [142].

6.2.4 *The ρ_S of Spearman and its connection with the PQD concept*

This coefficient can be defined as the linear correlation coefficient between the two uniform variables $U = F_1(X)$ and $V = F_2(Y)$, where F_1 and F_2 are the cumulative distribution function of respectively X and Y. Since U and V are uniform :

$$E(U) = E(V) = 1/2$$

and

$$var(U) = var(V) = 1/12.$$

Therefore:

$$\begin{aligned} \rho_S(X, Y) = corr(U, V) &= \frac{E(UV) - 1/4}{1/12} \\ &= 12.E(UV) - 3 \\ &= 12 \int_{[0,1]^2} UV du dv - 3 \\ &= 12\left(\int_{[0,1]^2} UV du dv - \int_{[0,1]} U du \int_{[0,1]} V dv\right). \end{aligned} \tag{6.3}$$

ρ_S is invariant to an increasing transformation of X and/or Y, it depends only on the ranks of the observations. $\rho_S/12$ is an average measure on the PQD property on the whole plane [160]. Indeed, using (7), we know that if $C(u, v)$ is the distribution function of the copula (U, V), then :

$$cov(U, V) = \int_{[0,1]^2} (C(u, v) - uv) du dv.$$

Therefore :

$$\rho_S = 12 \int_{[0,1]^2} (C(u, v) - uv) du dv$$

$$= 12 \int\int_{R^2} (F(x,y) - F_1(x)F_2(y))dF_1.dF_2. \quad (6.4)$$

The expression inside the parentheses corresponds to the evaluation of the PQD dependence at the point (x, y). ρ_S is equal to 1 when F attains the upper Fréchet bound, and to (-1) when it attains the lower Fréchet bound.

6.2.4.1 *A geometric interpretation of ρ_S*

Long and Krzysztofowicz [147] have shown that ρ_S is a measure of distance on the unit square which characterizes the density of the copula. Instead of considering only the squared distance d_1^2 from a point (U, V) to the main diagonal D_1 on the unit square as we did in the preceding section, they take :

$$\Lambda = d_2^2 - d_1^2$$

where

$$d_2^2 = (U + V - 1)^2/2$$

is the squared distance from (U, V) to the second diagonal $D_2 : v = 1 - u$. We have

$$\Lambda = (4UV - 2(U + V) + 1)/2.$$

Hence:

$$E(\Lambda) = 2E(UV) - 1/2$$

i.e.

$$\rho_S = 6E(\Lambda).$$

6.2.4.2 *Estimation of ρ_S*

From Eq. (6.4), we can derive an estimator of ρ_S given by (in obvious notation) :

$$\widehat{\rho_S} = 12 \sum_{i,j} (\widehat{F(i,j)} - \widehat{F_1(i)}.\widehat{F_2(j)})\hat{p}_i\hat{p}_j.$$

6.2.5 *Schweizer-Wolff's index of dependence*

An index closely related to Spearman's ρ_S is the index σ_{XY} introduced by Schweizer and Wolff [197]. Instead of considering the difference $C(u,v)-uv$ in formula 6.4, they use its absolute value to define :

$$\sigma_{XY} = 12 \int_{[0,1]^2} |C(u,v) - uv)| dudv$$

σ_{XY} is a measure of the volume between the surfaces $C(u,v)$ and $u.v$. Since $\int_{[0,1]^2} |\min(u,v) - uv| dudv = \frac{1}{12}$, we have the two equivalences :

$$\sigma_{XY} = 0 \Leftrightarrow (X,Y) \text{ independent}$$

$$\sigma_{XY} = 1 \Leftrightarrow X \text{ is a monotone function of } Y.$$

6.2.6 *The Kendall τ and its connection with LRD property*

Two independent pairs of variables (X,Y) and (X',Y'), with the same distribution $F(.,.)$ are called concordant if

$$P(X - X')(Y - Y') \geq 0$$

and discordant otherwise. The index τ is defined as :

$$\begin{aligned} \tau &= P[(X - X')(Y - Y') \geq 0] - P[(X - X')(Y - Y') < 0] \\ &= 2P[(X - X')(Y - Y') \geq 0] - 1. \end{aligned} \tag{6.5}$$

Since

$$\begin{aligned} P(X - X')(Y - Y') \geq 0) &= P(X > X', Y > Y') + P(X < X', Y < Y') \\ &= 2P(X > X', Y > Y'). \end{aligned}$$

If the pair (X,Y) is fixed at (x,y), so

$$P(X' < x, Y' < y) = F(x,y)$$

If now, (X,Y) varies over the whole plane, we have :

$$P(X' < X, Y' < Y) = \int F dF$$

and therefore :

$$\tau = 4\int F dF - 1. \tag{6.6}$$

In the particular case of Archimedean copulas with generator φ, using the results of section 4.6.6, this formula becomes :

$$\begin{aligned} \tau &= 4E(Z) - 1 \\ &= 4E(\lambda) + E(K(z)) \\ &= 4E(\lambda) + 1 \end{aligned} \tag{6.7}$$

where $Z = F(X,Y)$ has the cumulative function $K(z) = z - \lambda(z)$, and $\lambda(z) = \frac{\varphi(z)}{\varphi'(z)}$.

Nelsen [160] proved that $\frac{\tau}{2}$ represents an average measure of total positivity for the density f. He calculates :

$$T = \int_{-\infty}^{+\infty}\int_{-\infty}^{y_2}\int_{-\infty}^{+\infty}\int_{-\infty}^{x_2}[f(x_2,y_2)f(x_1,y_1) - f(x_2,y_1)f(x_1,y_2)]dx_1dy_1dx_2d$$

Let

$$T(x_2,y_2) = \int_{-\infty}^{y_2}\int_{-\infty}^{x_2}[f(x_2,y_2)f(x_1,y_1) - f(x_2,y_1)f(x_1,y_2)]dx_1dy_1$$

then

$$T(x_2,y_2) = f(x_2,y_2)F(x_2,y_2) - D_1F(x_2,y_2)D_2F(x_2,y_2).$$

Hence,

$$T = \int_{-\infty}^{+\infty}\int_{-\infty}^{+\infty}[f(x,y)F(x,y) - D_1F(x,y)D_2F(x,y)]dxdy.$$

From (6.6), the first term in the integral is $\frac{1}{4}(\tau+1)$, and the second is $\frac{1}{4}(1-\tau)$. Therefore $T = \frac{\tau}{2}$.

6.2.6.1 *Estimation of* τ

Given a sample of size n from the pair (X,Y), one can estimate τ by :

$$\hat{\tau} = \frac{\sharp(concordant\ pairs) - \sharp(discordant\ pairs)}{\binom{n}{2}}.$$

If we want to test the independence between X and Y, we need to compute the expectation and the variance of $\hat{\tau}$. Define the numerator K of $\hat{\tau}$ by : $K = \sum Z_{ii'}$ where:

$$Z_{ii'} = \begin{cases} 1 & \text{if } (X,Y)_i \text{ and } (X,Y)_{i'} \text{ concordant} \\ -1 & \text{otherwise.} \end{cases}$$

If the hypothesis of independence H_0 is valid, then $E(Z_{ii'}) = 0$ and $var(Z_{ii'}) = 1/2$. Consequently $E(K) = 0$ and $var(K) = E(K^2)$ $= E\left(\sum_{i=1}^{n-1} \sum_{i'=i+1}^{n} z_{ii'}\right)^2$. After some tedious but straightforward calculations, one finds :

$$var(K) = \frac{n(n-1)(2n+5)}{18}$$

hence

$$var(\hat{\tau}) = \frac{var(K)}{n(n-1)/2} = \frac{2(2n+5)}{9n(n-1)}.$$

For $n > 10$ the distribution of $\hat{\tau}$ is normal with a sufficient approximation and consequently one can presumably test the hypothesis of independence using the standard procedure. See, however, the warning by Hallin and Seoh (1998) [91].

6.2.7 *The Blomqvist medial coefficient*

This coefficient, also known as quadrant test of Blomqvist, evaluates the dependence at the "center" of a distribution. If X and Y are independent, then, in particular, $F(1/2, 1/2) = F_1(1/2)F_2(1/2) = 1/4$ The coefficient of Blomqvist is then defined as :

$$\beta = 4F(1/2, 1/2) - 1.$$

6.2.8 τ, ρ_S, β *and ordering on the distributions*

Lehmann (1963) [142] and later Tchen (1980) [216] have proved that Kendall's τ, Spearman's ρ_S, and Blomqvist's β are monotone functions of the underlying bivariate distributions F. If two bivariate distributions G and H have the same marginals, and if $G(x, y) \leq H(x, y)$ for all x and y, then the values of the three coefficients are smaller for G than for H.

6.2.9 *Constructing other global measures*

We note that there exists a multitude of measures in the literature related to the linear correlation coefficient ρ, Spearman's ρ_S and Kendall's τ which we have already discussed. Many non-parametric measures of dependence are based on the distance between a copula to the independent copula or to maximal dependent copula. For example the measure of the type :

$$\frac{F(x,y) - F_1(x).F_2(y)}{min(F_1(x), F_2(y)) - F_1(x)F_2(y)}$$

with or without normalization, with or without various averaging over the range of the distribution (either the joint or the product of the marginals). These are given by Schweiwer and Wolff (1981) [197], Tuncer and Sungur (1980) [220], Reimann (1987) [178], and earlier by Hoeffding (1948) [97]. Properties of these measures are closely related to Spearman's ρ_S and Kendall's τ.

6.2.10 *Indices for more than two variables*

Wolff [225] has extended the indices σ_{XY} and ρ_s to measure the dependence within the vector $\mathbf{X} = (X_1, ..., X_n)$ $(n > 2)$ with the distribution F. Let C be the copula associated to F. On the unit n-cube $I^n = [0,1]^n$, let us consider the two copulas corresponding to the independence and to the maximal dependence (the upper Fréchet bound):

$$C^0(u_1, ..., u_n) = \Pi_{i=1}^n u_i$$

and

$$C^+(u_1, ..., u_n) = Min(u_1, ..., u_n)$$

and the function corresponding to the lower Fréchet bound:

$$C^-(u_1, ..., u_n) = Max(u_1 + ... + u_n - n + 1, 0).$$

Let a_n and b_n be the volumes :

$$a_n = \int_{I_n} (Min(u_1, ..., u_n) - \Pi_{i=1}^n u_i) du_1 ... du_n$$

and

$$b_n = \int_{I_n} (\Pi_{i=1}^n u_i - Max(u_1 + ... + u_n - n + 1, 0)) du_1 ... du_n.$$

Then the indices σ_n and ρ_{nS} are defined as:

$$\sigma_n = a_n^{-1} \int_{I_n} |C(u_1, ..., u_n) - C^0(u_1, ..., u_n)| du_1 ... du_n$$

and

$$\rho_{nS} = a_n^{-1} \int_{I_n} (C(u_1, ..., u_n) - C^0(u_1, ..., u_n)) du_1 ... du_n.$$

The index σ_n satisfies all the axioms of Renyi (adapted to the case when $n > 2$). The index ρ_{nS} can be negative and does not satisfy the axiom on independence. However, if n is large the functions C^- and C^0 are not "far" and :

$$-\frac{b_n}{a_n} \leq \rho_{nS} \leq 1$$

where the quantity $-\frac{b_n}{a_n}$ is $0(\frac{n}{2^n})$ and tends to zero when n increases.

In the same way Kendall's τ can be extended to :

$$\tau_n = \frac{1}{2^{n-1} - 1} \left(2^n \int_{I_n} (C_n(u_1, u_2, ..., u_n) dC_n(u_1, u_2, ..., u_n) - 1 \right)$$

(Jouini and Clemen [120]). Again, one sees that the lower value of $\tau_n = -\frac{1}{2^{n-1}-1}$ tends rapidly to zero when n increases.

6.2.11 *Mutual information, relative entropy and derivatives measures*

6.2.11.1 *Definitions*

(1) If X is a random variable, with density $f_1(x)$, then the entropy or the measure of uncertainty is defined by :

$$E_X = - \int f_1(x) \log f_1(x) dx.$$

If we consider all the distributions with the density f defined on a compact set A of R, i.e. $\int_A f dx = 1$ and $f = 0$ on $R - A$, then the entropy is maximal when f is the uniform density on A [109].

(2) If (X, Y) is a pair of random variables with the density $f(x, y)$, and the marginal densities $f_1(x)$ and $f_2(y)$, then the entropy for this pair is :

$$E_{X,Y} = -\int\int f(x,y)\log(f(x,y))dxdy.$$

This entropy is maximum, when X and Y are independent. This definition can be generalized with an n-vector $(X_1, ..., X_n)$ in place of (X, Y).

(3) One can show [134] that

$$-\int\int f(x,y)\log(f(x,y))dxdy \leq -\int\int f(x,y)\log(f_1(x).f_2(y))dx.d$$

The mutual information [20] or relative entropy [109], [110], [121] is then defined as the difference between these two quantities :

$$\delta_{X,Y} = \int\int f(x,y)\log(\frac{f(x,y)}{f_1(x)f_2(y)}).dxdy. \tag{6.8}$$

If the components of (X, Y) are independent, then $\delta_{X,Y}$ is zero, and conversely when the dependence is maximal, $\delta_{X,Y}$ approaches infinity. To normalize this index, Joe [110] defines :

$$\delta^* = (1 - \exp(-2\delta))^{\frac{1}{2}}. \tag{6.9}$$

The index δ^* is confined to the interval $[0, 1]$, and in the case when the pair (X, Y) is bivariate normal is equal to the absolute value of the linear correlation coefficient $|\rho|$. Indeed, given

$$(X, Y)' \sim N(\mu_1, \mu_2, \Sigma)$$

where

$$\Sigma = \begin{bmatrix} \sigma_1^2 & \rho\sigma_1\sigma_2 \\ \rho\sigma_1\sigma_2 & \sigma_2^2 \end{bmatrix},$$

we have $\delta = -\frac{1}{2}.\log\frac{|\Sigma|}{\sigma_1^2\sigma_2^2}$ and consequently $\delta^* = |\rho|$.

6.2.11.2 *Examples*

(1) Generally, if X is a p-random vector of regressors and Y is the vector of the dependent variable, so that

$$(Y, X)' \sim N(\mu_Y, \mu_X, \Sigma)$$

with

$$\Sigma = \begin{bmatrix} \sigma_Y^2 & \Sigma_{YX} \\ \Sigma_{XY} & \Sigma_{XX} \end{bmatrix},$$

then

$$\delta^*_{YX} = \left[\frac{\Sigma_{YX}\Sigma_{XX}^{-1}\Sigma_{XY}}{\sigma_Y^2}\right]^{1/2}.$$

(2) If (X, Y) is a $p+q$-random normal vector with the covariance matrix:

$$\Sigma = \begin{bmatrix} \Sigma_{XX} & \Sigma_{XY} \\ \Sigma_{YX} & \Sigma_{YY} \end{bmatrix}$$

then, $\delta = -\frac{1}{2}.\log\frac{|\Sigma|}{|\Sigma_{XX}||\Sigma_{YY}|}$. However one can expand (see e.g. Searle [199] page 43) the determinant of this block-partitioned matrix as follows :

$$\begin{aligned} |\Sigma| &= |\Sigma_{YY}||\Sigma_{XX} - \Sigma_{XY}\Sigma_{YY}^{-1}\Sigma_{YX}| \\ &= |\Sigma_{YY}||\Sigma_{XX}||I - \Sigma_{XX}^{-1}\Sigma_{XY}\Sigma_{YY}^{-1}\Sigma_{YX}|. \end{aligned}$$

Hence

$$\delta = -\frac{1}{2}.\log|I - \Sigma_{XX}^{-1}\Sigma_{XY}\Sigma_{YY}^{-1}\Sigma_{YX}|.$$

If λ_i is a strictly positive eigenvalue of $\Sigma_{XX}^{-1}\Sigma_{XY}\Sigma_{YY}^{-1}\Sigma_{YX}$ (it is also less than 1), then $1-\lambda_i$ is a non-zero eigenvalue of $I - \Sigma_{XX}^{-1}\Sigma_{XY}\Sigma_{YY}^{-1}\Sigma_{YX}$ and therefore its determinant is :

$$\Pi_{i=1}^{s}(1-\lambda_i).$$

Hence

$$\delta^* = \left[1 - \Pi_{i=1}^{s}(1-\lambda_i)\right]^{1/2},$$

where s is the number of non-zero eigenvalues. In fact $\lambda_i = \rho_i^2$ is a canonical correlation coefficient (see e.g., Johnson and Wichern

[117], and the preceding formula indicates that the normalization of the relative entropy depends only on the non-zero eigenvalues $\{\lambda_1, ..., \lambda_s\}$. Lin [145] has shown that it is a property of functions which are invariant with respect to a linear transformation (location and scale changes) of X and Y.

(3) Bivariate Pareto distribution
We can calculate the entropy for the pair (X, Y). For X, we have :

$$\begin{aligned} E_X &= -\int_{\theta_1}^{\infty} f_1(x) \log f_1(x) dx \\ &= -\int_{\theta_1}^{\infty} f_1(x)[\log a + a \log \theta_1 - (a+1). \log(x)]\, dx \\ &= -\log(a) - a \log \theta_1 + (a+1) \int_{\theta_1}^{\infty} f_1(x) dx \\ &= -\log a - a \log \theta_1 + (a+1)[\frac{1}{a} + \log \theta_1] \\ &= \log \theta_1 - \log a + 1 + \frac{1}{a}. \end{aligned} \tag{6.10}$$

Similarly:

$$E_{X,Y} = -[\log a + \log(a+1) + \log \theta_1 + \log \theta_2 + (a+2)[\frac{1}{a} + \frac{1}{a+1}].$$

Hence:

$$\delta_{X,Y} = \log \frac{a+1}{a} - (a+2)[\frac{1}{a} + \frac{1}{a+1}] + 2(1 + \frac{1}{a}).$$

Therefore $\delta_{X,Y}$ is defined for $a > 0$, over a larger domain than ρ.

(4) Bivariate Inverted Dirichlet Distribution
Tiao and Guttman [218] have constructed a multivariate inverted Dirichlet distribution, whose density for the case of two variables is given by :

$$f(x, y) = \frac{\Gamma(m_1 + m_2 + m_3)}{\Gamma(m_1)\Gamma(m_2)\Gamma(m_3)} \cdot \frac{x^{m_1 - 1} y^{m_2 - 1}}{(1 + x + y)^{m_1 + m_2 + m_3}}, \ x \geq 0, \ y \geq 0. \tag{6.11}$$

Kapur and Dhande [121] have shown that if $m_1 = m_2 = 1$ and $m_3 = a$, and if one transforms $U = \theta_1(X + 1)$ and $V = \theta_2(Y + 1)$, then the pair (U, V) has a bivariate Pareto distribution when (X, Y)

has a bivariate inverted Dirichlet distribution. The coefficient of correlation is :

$$\rho(x,y) = \frac{(m_1 m_2)^{1/2}}{(m_3 + m_1 - 1)^{1/2}(m_3 + m_2 - 1)^{1/2}}$$

and the relative entropy is given by [121] :

$$\begin{aligned} \delta_{X,Y} &= \log(\frac{\Gamma(m_1+m_2+m_3)\Gamma^3(m_3)}{\Gamma(m_1+m_3)\Gamma(m_2+m_3)}) \\ &+ \sum_{i=1}^{i=2}(m_i+m_3)\frac{\Gamma'(m_i+m_3)}{\Gamma(m_i+m_3)} \\ &-(m_1+m_2+m_3)\frac{\Gamma'(m_1+m_2+m_3)}{\Gamma(m_1+m_2+m_3)} - m_3\frac{\Gamma'(m_3)}{\Gamma(m_3)}. \end{aligned} \quad (6.12)$$

In the special case when $m_1 = m_2 = m$ and $m_3 = a$, we have :

$$\rho(x,y) = \frac{m}{m+a-1}$$

and

$$\begin{aligned} \delta_{X,Y} &= \log(\frac{\Gamma(2m+a)\Gamma(a)}{\Gamma^2(m+a)}) + 2(m+a)\frac{\Gamma'(m+a)}{\Gamma(m+a)} \\ &-(2m+a)\frac{\Gamma'(2m+a)}{\Gamma(2m+a)} - a\frac{\Gamma'(a)}{\Gamma(a)}. \end{aligned}$$

6.2.11.3 *Lin's measure of association*

Since the measures which are invariant with respect to a linear transformation depend only on the $\{\lambda_i \; i = 1, ..., s\}$, the non-zero eigenvalues of $\Sigma_{XX}^{-1}\Sigma_{XY}\Sigma_{YY}^{-1}\Sigma_{YX}$, Lin [145] proposed to construct a measure of association between X and Y as a function of these eigenvalues.

Definition : Let f be a differentiable and strictly increasing function, $f : [0,1] \mapsto [0,1]$ with $f(0) = 0$ and $f(1) = 1$ and let g be a differentiable and monotone function $g : [0,1] \mapsto [a,b]$, with $0 \le a < b \le \infty$ and either :

- $g(0) = a$ and $g(1) = b$
- or $g(0) = b$ and $g(1) = a$.

Then the measure of association is defined to be

$$\rho_{f,g}(X,Y) = g^{-1}\left\{\sum_{i=1}^{s} c_i g[f(\lambda_i)]\right\}$$

if $s > 0$, and otherwise $\rho_{f,g}(X,Y) = 0$, with the constraints $c_i \geq 0$ and $\sum_{i=1}^{s} c_i = 1$.

But if $a = 0$ and $b = \infty$, the condition $\sum_{i=1}^{s} c_i = 1$ is not required; however at least one of the c_i, $i = 1, ..., s$, must be positive.

Any $\rho_{f,g}$ defined as above possesses the following properties.

(1) Symmetry : $\rho_{f,g} = \rho_{g,f}$.
(2) $0 \leq \rho_{f,g} \leq 1$. Indeed, the eigenvalues λ_i are between zero and one.
(3) $\rho_{f,g} = 0 \Leftrightarrow \Sigma_{XY} = 0$, therefore in the case where X and Y are normal vectors this is equivalent to (X,Y) being independent.
(4) If $Y = HX$ where H is a $p \times q$ matrix of rank p, then $\rho_{f,g} = 1$.

Example : The measure $\delta^*_{X,Y}$ defined by Joe (Eq. 6.9) is a ρ_{fg} measure with $f = I$ and $g(y) = -\log(1-y)$, and $c_i = 1$ for all i. In that case $\delta^*_{X,Y} = 1$ implies that there exists a matrix H of rank p such that $Y = HX$.

6.2.11.4 *Zografos's measure of association*

Starting from the Lin's measure of dependence, Zografos [233] proposed a new measure of dependence $\eta_{X,Y}$ between the two random vectors X and Y equal to $\rho_{f,g}(\mathcal{S}_{\mathcal{X}}, \mathcal{S}_{\mathcal{Y}})$, where $\mathcal{S}_{\mathcal{X}}$ and $\mathcal{S}_{\mathcal{Y}}$ are the score function vectors:

$$\mathcal{S}_X = \frac{\partial \log f(x,y)}{\partial x} = \left(\frac{\partial \log f(x,y)}{\partial x_1}, ..., \frac{\partial \log f(x,y)}{\partial x_p} \right)'_{1p}$$

$$\mathcal{S}_Y = \frac{\partial \log f(x,y)}{\partial y} = \left(\frac{\partial \log f(x,y)}{\partial y_1}, ..., \frac{\partial \log f(x,y)}{\partial y_q} \right)'_{1q}.$$

Let $f(x,y)$ satisfy certain regularity conditions established by Papathanassiou [164] namely :

(1) continuous partial derivatives:
$\frac{\partial \log f(x,y)}{\partial x_i}$ $\forall\, i = 1, ..., p$, $\frac{\partial \log f(x,y)}{\partial y_j}$ $\forall\, j = 1, ..., q$, $\frac{\partial^2 \log f(x,y)}{\partial x_i \partial y_j}$ $\forall$ *i and j* are continuous.
(2) vanishing on the boundaries of the domain of definition:
f is defined on a set E, which is a union of open connected sets in R^{p+q} , and $f(x,y)$ tends to zero monotonically as x or y approaches the boundary of E along the coordinates axis.
(3) The Fisher's type positive semi-definite block symmetric information matrix is non-singular :

$$I_f = \begin{bmatrix} I_{11} & I_{12} \\ I_{21} & I_{22} \end{bmatrix},$$

where

$$I_{11} = E\left(\frac{\partial \log f(x,y)}{\partial x_i}\frac{\partial \log f(x,y)}{\partial x_j}\right) \quad i,\ j\ = 1, ..., p,$$

$$I_{22} = E\left(\frac{\partial \log f(x,y)}{\partial y_i}\frac{\partial \log f(x,y)}{\partial y_j}\right) \quad i,\ j\ = 1, ..., q$$

$$I_{12} = E\left(\frac{\partial \log f(x,y)}{\partial x_i}\frac{\partial \log f(x,y)}{\partial y_j}\right) i = 1, ..., p,\ j = 1, ..., q.$$

In that case $E(S_X) = E(S_Y) = 0$ and $I_{11} = var(S_X)$, $I_{22} = var(S_Y)$ and $I_{12} = cov(S_X, S_Y)$.

Since $\mathcal{S}_\mathcal{X}$ and $\mathcal{S}_\mathcal{Y}$ play the role of X and Y in the definition of ρ_{fg}, the matrix $\Sigma_{XX}^{-1}\Sigma_{XY}\Sigma_{YY}^{-1}\Sigma_{YX}$ becomes $I_{11}^{-1}I_{12}I_{22}^{-1}I_{21}$ and the λ_i are the eigenvalues of the latter matrix.

The measure $\eta_{X,Y}$ has the same properties as $\rho_{f,g}$:

- symmetry
- $0 \leq \eta_{X,Y} \leq 1$
- $\eta_{X,Y} = 0 \Leftrightarrow (X,Y)$ are independent. Indeed using property 3 of Section 6.2.11.3 :

$$\eta_{X,Y} = 0 \Leftrightarrow I_{12} = 0.$$

This is, however, equivalent to

$$\left(\frac{\partial \log f(x,y)}{\partial x_i}.\frac{\partial \log f(x,y)}{\partial y_j}\right) = 0\ \forall\, i,\ \forall\, j.$$

Also since the differentiation under the sign integral is justified, this is equivalent to :

$$-\int\int \frac{\partial^2 \log f(x,y)}{\partial x_i \partial y_j} f(x,y)dxdy = 0, \forall\, i\ , \forall\, j$$

and hence $\frac{\partial^2 \log f(x,y)}{\partial x_i \partial y_j} = 0$ *a.e.* in x *and* y. The last relation is thus equivalent to the independence of X and Y.

- The measure is invariant under a linear transformation.

- If $p = q$ and $Y = AX$ where A is a non-singular matrix, then $\eta_{X,Y} = 0$ (see property 4 in Section 6.2.11.3).

Examples :

(1) Let (X, Y) be a bivariate random vector:
Then I_f becomes :

$$I_f = \Sigma^{-1} = \frac{1}{1-\rho^2} \cdot \begin{bmatrix} \frac{1}{\sigma_X^2} & \frac{-\rho}{\sigma_X \sigma_Y} \\ \frac{-\rho}{\sigma_X \sigma_Y} & \frac{1}{\sigma_Y^2} \end{bmatrix}.$$

The eigenvalue of $I_{11}^{-1} I_{12} I_{22}^{-1} I_{21}$ is ρ^2. With $g = Id$, and $f(x) = \sqrt{x}$, one obtains $\eta_{X,Y} = |\rho|$, which is δ^*, the measure of Joe.

(2) (X, Y) is a multivariate random vector:
In that case, using a theorem of Mayer-Wolf [153], we obtain :

$$I_f = \Sigma^{-1}$$

and from Anderson [6] we have:

$$I_{11}^{-1} I_{12} I_{22}^{-1} I_{21} = \Sigma_{XY} \Sigma_{YY}^{-1} \Sigma_{YX} \Sigma_{XX}^{-1}.$$

These two products of matrices have therefore the same eigenvalues λ_i . If we take $f = Id$ and $g(y) = -\log(1-y)$, and $c_i = 1$ then :

$$\eta_{X,Y} = 1 - \Pi_{i=1}^{s}(1 - \lambda_i)$$

i.e $\eta_{X,Y}$ is the square of the Joe measure $\delta^*_{X,Y}$.

(3) The bivariate inverted Dirichlet distribution :
In the particular case when $m_1 = m_2 = m$ and $m_3 = a$, Zografos [233] has shown that :

$$I_f = \frac{a(a+1)}{2m+a+1} \cdot \begin{bmatrix} \frac{m+a+3}{m-2} & -1 \\ -1 & \frac{m+a+3}{m-2} \end{bmatrix}.$$

With $f = g = Id$, $c_1 = 0$ and $c_2 = 1$, $\eta_{X,Y} = (\frac{m-2}{m+a+3})^2$ which is a simpler measure that the relative entropy suggested by Kapur and Dhande [121] .

6.3 Local Indices of Dependence

6.3.1 *Motivation*

We have seen that ρ_S is an average measure of the PQD dependence. . The following example presented in Kotz *et al.* [133] shows that a distribution with a high ρ_S may not be PQD.

Y/X	1	2	3	
1	0.25	0.10	0.01	0.36
2	0.01	0.03	0.22	0.26
3	0.02	0.36	0.10	0.38
	0.28	0.39	0.33	

Here one can estimate ρ_S by :

$$\widehat{\rho_S} = 12 \sum_{i,j} (\widehat{F(i,j)} - \widehat{F_1(i)}.\widehat{F_2(j)})\hat{p}_i \hat{p}_j$$

we obtain :

$$\widehat{\rho_S} = 0.53.$$

However the distribution is not PQD. Indeed: $F(2,2) = 0.39$ is less than $F_1(2)F_2(2) = 0.4154$.

6.3.2 *Local definition of the dependence*

Our aim is to define local dependence and "remaining" dependence. (We use this last term to indicate dependence on a part of R^2, specifically in case of survival variables in an orthant beyond a certain point. If $V(x_0, y_0)$ is an open neighborhood of (x_0, y_0), then a distribution $F(x, y)$ is PQD in the neighborhood $V(x_0, y_0)$ provided

$$\forall (x, y) \in V(x_0, y_0) \text{ we have } S(x, y) \geq S(x)S(y).$$

If

$$V(x_0, y_0) =]x_0, +\infty[\times]y_0, +\infty[$$

we then arrive at the remaining PQD. In the same manner a local or remaining LRD can be defined.

Example : the Cauchy distribution is PQD in the first and third quadrant and NQD (negative quadrant dependence) in the others.

6.3.3 *Local ρ_S and τ*

We can restrict ρ_S and τ to an open neighborhood of (x_0, y_0), and define:

$$\rho_{S,(x_0,y_0)} = \frac{12\int\int_{V(x_0,y_0)}(C(u,v) - uv)dudv}{\int\int_{V(x_0,y_0)} dudv}$$

and

$$\tau_{(x_0,y_0)} = \frac{4\int\int_{V(x_0,y_0)} C(u,v)dC - 1}{\int\int_{V(x_0,y_0)} dC}.$$

In the case when $V(x_0, y_0) =]x_0, +\infty] \times]y_0, +\infty[$, it is straightforward to estimate $\tau_{(x_0,y_0)}$ by counting the remaining concordant and discordant pairs, and to estimate the variance of this estimator from n_0, the number of remaining observations.

6.3.4 *Local correlation coefficient of Bjerve and Doksum*

6.3.4.1 *Motivation and historical remarks*

Since K.Pearson's paper (1905), many authors described situations where the linear correlation coefficient takes unsuccessfully into account the association between two random variables X and Y . The reason for this is that either the dependence does not exist on the whole plane or it is not linear, or the variance of Y given $X = x$ is not constant (heteroscedasticity). Blyth [25] reproduces in his paper an example extracted from "On the general theory of skew correlation and non-linear regression" (1905), where K. Pearson analyses a data set, consisting of measurements of auricular heights of 2272 schools girls according to their ages (between 2 and 23). In this example, Pearson proposes to fit the data by a cubic regression curve :

$$\hat{\mu} * (x) = 0.28 + 0.723.x_* - 0,0296.x_*^2 - 0,00222x_*^3$$

where x is the age and y the auricular height, * indicate that the data are centered around the empirical mean, and $\mu(x)$ is $E(Y/X = x)$. Pearson also plots the empirical conditional standard deviations, scaled by $\hat{\sigma}_Y$, that

Blyth reconstructs them to be :

$$\hat{\sigma}(x) = 4.5 + 0.17x$$

where $\sigma^2(x) = var(Y/X = x)$.

6.3.4.2 *Definition, properties and limits*

(1) Definition: If (X, Y) is bivariate normal $N(\mu_X, \mu_Y, \sigma_X, \sigma_Y)$ the regression curve is simply

$$E(Y/X = x) = \alpha + \beta.x + \epsilon\ ,$$

with $var(\epsilon) = \sigma^2$.
The linear correlation coefficient is then :

$$\rho = \frac{\sigma_X \beta}{(\sigma_X \beta)^2 + \sigma^2}.$$

If (X, Y) is non-normal, a local extension of ρ is defined by :

$$\rho(x) = \frac{\sigma_X \beta(x)}{(\sigma_X \beta(x))^2 + \sigma(x)^2}$$

where $\beta(x) = \frac{\partial \mu(x)}{\partial x}$, (assuming differentiability of $\mu(x)$).

(2) $\rho(x)$ is standardized, that is

$$-1 \leq \rho(x) \leq 1.$$

(3) The components of (X, Y) being independent implies that $\rho(x) \equiv 0\ \forall\ x$.

(4) $\rho(x) \pm 1$ for almost all x is equivalent to Y being a function of X.

(5) In general $\rho(x)$ is not symmetric, but it is possible to construct a symmetrized version $\tilde{\rho}$ by $\tilde{\rho(x)} = sign(\rho(x))\sqrt{\rho(x).\rho(y)}$ if $\rho(x)$ and $\rho(y)$ have the same sign and $\tilde{\rho(x)} = 0$ otherwise.

(6) $\rho(x)$ is scale-free but not marginal-free, that is linear transformations of X and Y, $X* = aX + b$ and $Y* = cY + d$, with c and d having the same sign, leaves $\rho(x)$ unchanged, but the transformations $U = F_1(X)$ and $V = F_2(Y)$ result in $\rho(u)$ which is different from $\rho(x)$.

(7) the PRD (positive regression dependence) and $\rho(x)$:
If $\rho(x) \geq 0$ then PRD is valid. We can also define a local PRD when $\rho(x)$ is positive in a neighborhood of (x_0, y_0).

6.3.4.3 *Estimations and properties of the estimators*

Bjerve and Doksum [21] use a non-parametric estimation method for the parameters $\mu(x) = E(Y/X = x)$, $\beta(x) = \frac{d\mu(x)}{dx}$ and $\sigma(x) = \sqrt{var(Y/X = x)}$.

Let $K(u)$ be the Epanechnikov kernel, $K(u) = 0.75(1 - u^2)1_{|u|\leq 1}(u)$, and x_0 be a point of a subdivision on the x-axis.

$y = a(x_0) + b(x_0)x$ is a weighted least square line computed from the sample $\{(x_i, Y_i)\ i = 1, ..., n\}$, with the weight $w_i = K((x_i - x_0)/h)$, where $h = s_X$ is the empirical standard deviation of $\{x_i,\ i = 1, ..., n\}$. The estimates are therefore :

$$\hat{\mu}(x_0) = a(x_0) + b(x_0)x_0$$

and

$$\hat{\beta}(x_0) = b(x_0).$$

Similarly to estimate $\sigma^2(x_0)$, consider the sample $\{(x_i, \hat{\epsilon}_i^2),\ i = 1, ..., n)\}$, where $\hat{\epsilon}_i = Y_i - \hat{\mu}(x_i)$ is the i-th residual, and fit the line $y = c(x_0) + d(x_0)x$ with the same weights w_i as above. The estimate of $\sigma^2(x_0)$ is then $c(x_0) + d(x_0)x_0$ and the estimate of the local correlation coefficient is

$$\hat{\rho}(x_0) = \frac{s_X\hat{\beta}(x_0)}{\sqrt{s_X^2\hat{\beta}^2(x_0) + \hat{\sigma}^2(x_0)}}.$$

The above procedure is then repeated for all the points of the subdivision.

The delta-method allows us to estimate the standard deviation of $\widehat{\rho(x)}$. One expands

$$h(s,t) = \frac{s}{\sqrt{s^2 + t}}$$

around the values $s_0 = \sigma_x\beta(x_i)$ and $t_0 = \sigma^2(x_i)$, and uses the fact that $\beta(x_i)$ and $\sigma(x_i)$ are independent. In a subsequent paper Doksum *et al.* [61] propose to use two other non-parametric methods to estimate $\hat{\rho}(x)$:the nearest-neighbor method method based on the Gasser–Müller [79] kernel. They provide conditions for consistency and asymptotic normality of $\hat{\rho}(x)$. They choose an optimal bandwidth by using a parametric model and searching for the optimal bandwidth under this model.

6.3.5 *Correlation ratio*

Renyi (1959) [179] motivated by Kolmogorov defines the correlation ratio :

$$K_Y(X) = \frac{Var(E(Y/X))}{Var(Y)}.$$

When the distribution of (X, Y) is standardized bivariate normal, $K_X(Y) = \rho$. One can show that $K_Y(X) = sup_g|\rho(Y, g(X))|$ for all measurable functions g, such that $g(X)$ exists. One also deduces that :

$$K_Y(X) = |\rho(Y, g(X))| \text{ iff } g(X) = aE(Y/X) + b \quad (a.e).$$

6.3.6 *Local dependence function of Bairamov and Kotz*

Bairamov *et al.* [13] have defined a "local dependence function" $H(x, y)$, which provides a local point of view on dependence at a point (x, y). If $\mu(y) = E(X/Y = y)$ and $\mu(x) = E(Y/X = x)$, then :

$$H(x, y) = \frac{E(X - \mu(y))(Y - \mu(x))}{\sqrt{E(X - \mu(y))^2 . E(Y - \mu(x))^2}}.$$

$H(x, y)$ is obtained from the expression of the linear correlation coefficient by replacing expectations $E(X)$ and $E(Y)$ by the conditional expectations $\mu(y) = E(X/Y = y)$ and $\mu(x) = E(Y/X = x)$ respectively.

After some algebraic manipulations the expression can be rewritten as:

$$H(x, y) = \frac{\rho + \varphi_X(y)\varphi_Y(x)}{\sqrt{(1 + \varphi_X^2(y))(1 + \varphi_Y^2(x))}} \tag{6.13}$$

where $\varphi_X(y) = \frac{E(X)-\mu(y)}{\sqrt{Var(X)}}$ and $\varphi_Y(x) = \frac{E(Y)-\mu(x)}{\sqrt{Var(Y)}}$.

For a further analysis of Equation (6.13), let $a \le \varphi_X(y) \le b$ and $a \le \varphi_Y(x) \le b$ (possible including $a = -\infty$ and $b = \infty$ for all $(x, y) \in D_{XY}$ where D_{XY} denotes the support of (x, y)). Consider the function :

$$h(t, z) = \frac{\rho + tz}{\sqrt{(1 + t^2)(1 + z^2)}}.$$

The partial derivative of $h(t, z)$ w.r.t. z is :

$$\frac{\partial h}{\partial z} = \frac{t - z\rho}{(1 + z^2)^{\frac{3}{2}}\sqrt{1 + t^2}}$$

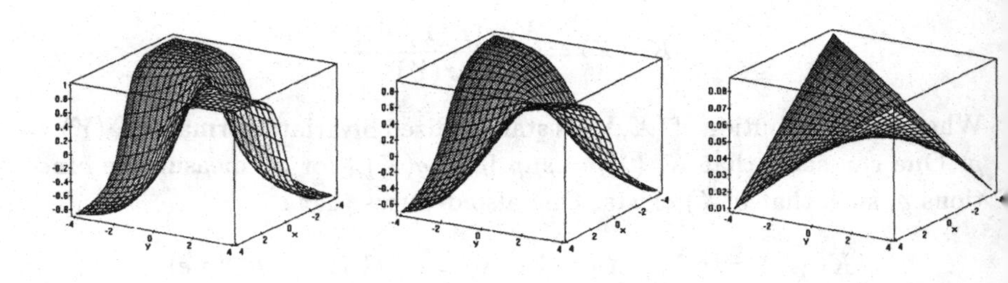

Fig. 6.3 Graph of $H(x,y)$ corresponding to standardized normal distributions with $\rho = 0.95$, $\rho = 0.5$, $\rho = 0.005$.

and the other partial derivative is obtained by exchanging z and t. These partial derivatives are zero at the point $t = 0$ and $z = 0$. Calculating the second derivatives at the point $(0,0)$ we find :

$$\left(\frac{\partial^2 h}{\partial z^2}\frac{\partial^2 h}{\partial t^2} - (\frac{\partial h}{\partial t \partial z})^2\right)(0,0) = \rho^2 - 1$$

which is negative if $|\rho| < 1$. Therefore $h(t,z)$ has a saddle point at the point (0,0), that is $H(x,y)$ has a saddle point at the point (x^*,y^*) satisfying: $\varphi_X(y^*) = 0$, i.e. $E(X) = \mu(y^*)$ and, $\varphi_Y(x^*) = 0$, namely $E(Y) = \mu(x^*)$ and at this point $H(x^*,y^*) = \rho$.

It is easy to see that $h(t,z)$ has a maximum at the boundary points (a,a) and (b,b) and a minimum at (a,b) and (b,a). The local dependence function has the following properties :

(1) If (X,Y) are independent, then $H(x,y) = 0$ for all (x,y) .
(2) $|H(x,y)| \le 1$ for all (x,y).
(3) If $|H(x,y)| = 1$ for some (x,y), then $\rho \neq 0$.
(4) Let $\mu(x)$ and $\mu(y)$ be differentiable functions of their arguments. If $H(x,y) = 0$ for all (x,y), then either $\mu(x)$ or $\mu(y)$, or both are constant.

6.3.7 *Measures of the tail dependence*

Coles *et al.* [50] have proposed two indices to measure tail dependence (see Chapter 3, Section 3.2.10) and a diagnosis of such a dependence. Recall that for a copula $C(u, v)$, the upper tail dependence is defined as :

$$\delta = lim_{u\to 1-}\frac{\overline{C}(u,u)}{1-u}.$$

This formula can be rewritten as :

$$\begin{aligned}\delta &= lim_{u\to 1-}\frac{1-2u+C(u,u)}{1-u}\\ &= 2-lim_{u\to 1-}\frac{1-C(u,u)}{1-u}\\ &\sim 2-lim_{u\to 1-}\frac{\log C(u,u)}{\log u}.\end{aligned} \tag{6.14}$$

Coles *et al.* check the behaviour of the function $\delta(u) = 2 - \frac{\log C(u,u)}{\log u}$ to diagnose tail dependence. ($\delta = lim_{u\to 1-}\delta(u)$.)

For independent variables $\delta(u) = 0$, for the Fréchet lower bound $C^{-}(u, u) = max(0, 2u - 1)$, $\delta(u) = 1$, and for Gumbel-Hougaard family $\delta(u) = 2 - 2^{\alpha}$. More generally $\delta(u)$ is constant (different from zero) for bivariate extreme value distributions. For asymptotically independent variables (for example for the normal copula model) $\delta(u)$ may be a complicated function of u, which tends to zero when u tends to one, and tends very slowly (and ultimately abruptly) if the correlation between U and V is high. So an estimation of $\delta(u)$ may be a bad indicator of tail dependence since the empirical observations are taken when $u < 1$.

To complete the diagnosis, Cole *et al.* propose to use another function:

$$\overline{\delta(u)} = \frac{2log(1-u)}{log\overline{C}(u,u)} - 1$$

which is in the range of $[-1, +1]$.

For asymptotically dependent variables the limit $\overline{\delta(u)}$ when u tends to one is $\overline{\delta} = 1$. For independent variables $\overline{\delta}(u) = 0$ and therefore $\overline{\delta} = 0$. For asymptotically independent variables $\overline{\delta}$ measures the strength of dependence, for example for the normal copula model $\overline{\delta}$ is equal to ρ, the correlation coefficient.

The two indices $(\delta, \overline{\delta})$ can therefore be used to identify extremal dependence: $(\delta > 0, \overline{\delta} = 1)$ signifies asymptotic dependence in which case the value of δ determines a measure of tail dependence; ($\delta = 0, \overline{\delta} < 1$) signifies asymptotic independence, in which case the value of $\overline{\delta}$ determines the strength of dependence.

6.3.8 *Several local indices applicable in survival analysis*

In the field of survival analysis, there is a need for time-dependent measures of dependence, to identify, for example in medical studies, the time of maximal association between the interval from remission to relapse and the next interval from relapse to death, or to determine the genetic character of a disease by comparing the degree of association between the lifetimes of monozygotic or dizygotic twins [100].

6.3.8.1 *The covariance function of Prentice and Cai*

With the aim to propose an estimator of the bivariate survival function in case of censored data, Prentice and Cai [173] have defined a covariance function characterizing the dependence between two failure time variables independent of the marginal distributions. They were searching for measures that express the dependence between (X, Y) as a function of follow-up durations x and y. Let $N_1(x)$ and $N_2(y)$ be the couting processes ($N_1(x) = 1$ if $X \leq x$ and $N_1(x) = 0$ otherwise), $H_{01}(x)$ and $H_{02}(y)$ be the cumulative marginal hazards and $M_1(x) = N_1(x) - H_{01}(x \wedge X)$ and $M_2(y) = N_2(y) - H_{02}(y \wedge Y)$, the martingales arising in the decomposition of N_i, $i = 1, 2$, with respect to the filtration defined by X (or Y), then the covariance function is defined as :

$$C(x, y) = cov\left(M_1(x), M_2(y)\right) = E\left(M_1(x).M_2(y)\right).$$

$C(x, y)$ and the marginals allow us to retrieve the survival function. Prentice and Cai show that :

$$S(x, y) = S_1(x)S_2(y)\{1 + \int_0^x \int_0^y (S_1(x)S_2(y))^{-1} C(du, dv)\}.$$

For example with the exponential marginals $S_1(x) = exp(-x)$ and $S_2(y) = exp(-y)$ and with $C(du, dv) = \theta \exp(-2(u + v))dudv$, one obtains:

$$S(x, y) = exp(-(x + y))\{1 + \theta.(1 - exp(-x))(1 - exp(-y))\}$$

which is a bivariate exponential distribution considered by Gumbel [88].

Shih and Louis [206] have considered the estimator of this covariance function in case of frailty models to search for an early and late dependence for censored survival data on AIDS.

6.3.8.2 *The conditional covariance rate of Dabrowska*

Dabrowska *et al.* [53], in the case where (X, Y) are two survival variables, proposed to use "the conditional covariance rate" :

$$l(x,y) = D_1 D_2 Log(S) = h(x,y) - h_1(x,y)h_2(x,y)$$

or equivalently :

$$l(x,y) = -D_1 h_2(x,y) = -D_2 h_1(x,y)$$

where $h_1(x,y)$ and $h_2(x,y)$ are two conditional hazards (see Chapter 1).

(1) : Interpretation from the frailty model (Chapter 4, section 4.5.5): Starting from Equation (4.21) :

$$S(x,y) = \int\int exp(-w_1 H_{10}(x) + H_{20}(y))dG(w_1, w_2)$$

we obtain for the partial derivatives:

$$D_i S = -\int\int w_i h_{10}(x) exp(-w_1 H_{10}(x) + H_{20}(y))dG(w_1, w_2), \; i = 1, 2,$$

and for the conditional hazards :

$$\begin{aligned} h_i(x,y) &= -\frac{D_i S}{S} = h_{i0}(x) \frac{\int\int w_i exp(-w_1 H_{10}(x) + H_{20}(y))dG(w_1, w_2)}{\int\int exp(-w_1 H_{10}(x) + H_{20}(y))dG(w_1, w_2)} \\ &= h_{i0}(x) \frac{\int\int w_i P(X > x, Y > y)/w_1, w_2)dG(w_1, w_2)}{\int\int P(X > x, Y > y)/w_1, w_2)dG(w_1, w_2)} \\ &= h_{i0}(x) E(W_i / X > x, Y > y), \; i = 1, 2. \end{aligned} \tag{6.15}$$

Using the same procedure, we find for the bivariate hazard:

$$h(x,y) = h_{10}(x) h_{20}(y) E(W_1 W_2 / X > x, Y > y).$$

Therefore :

$$l(x,y) = h_{10}(x) h_{20}(y) cov(W_1, W_2 / X > x, Y > y).$$

An analogous proof is provided in Anderson *et al.* [6], but only for one frailty variable.

(2) Interpretation : A discretized version of l:
Suppose that x takes on the values $1, 2, ..., i, i+1, ..., k$ and y the values $1, 2, ..., j, j+1, ..., m$.
Denote $S_{i,j} = P(x \geq i, Y \geq j)$ and define

$$SS_{ij} = \frac{S_{i,j} S_{i+1,j+1}}{S_{i+1,j} S_{i,j+1}},$$

i.e. SS_{ij} is an odds-ratio among the four adjacent cells, which measures the interaction among these cells. Define now $LL_{ij} = log(SS_{ij})$. LL_{ij} is then a discretized version of $l(x,y)$. Indeed:

$$\lim_{dx \to 0, dy \to 0} Log\left(\frac{S_{x,y} S_{x+dx,y+dy}}{S_{x+dx,y} S_{x,y+dy}}\right) = D_1 D_2 Log(S).$$

(3) l and the joint distribution :
- The function $l(x,y)$ is symmetric.
- l and the two marginals completely determine the distribution. Indeed,

$$S(x,y) = S_1(x) S_2(y) \exp \int_0^x \int_0^y l(u,v)\, du\, dv$$

- l is not marginal-free : Indeed, if C represents the copula function associated with the given survival functions, and c is the Copula's density :

$$S(x,y) = C(S_1(x), S_2(y))$$

then the density function is :

$$f(x,y) = D_1 D_2 S(x,y) = c(S_1(x), S_2(y)) f_1(x) f_2(y)$$

and the bivariate and conditional hazards are :

$$h(x,y) = \frac{f(x,y)}{S(x,y)} = \frac{c(S_1(x), S_2(y)) f_1(x) f_2(y)}{C(S_1, S_{2)}}$$

$$h_i(x,y) = -\frac{D_i S}{S} = \frac{D_i C(S_1, S_2) f_i}{C(S_1, S_2)} \; \forall\, i = 1, 2.$$

Thus

$$l = h - h_1 h_2 = f_1 f_2 (\frac{c}{C} - \frac{D_1 C D_2 C}{C^2})$$

i.e.

$$l = f_1 g_2 l_c$$

where l_c represents the conditional covariance rate for the copula.

(4) l and positive dependence:
- If X and Y are independent, this implies at $l \equiv 0$ and vice-versa.
- The condition $l(x, y) \geq 0, \forall (x, y)$, corresponds to positive dependence. More precisely if $l(x, y) \geq 0$, then $h_2(x, y)$ is a decreasing function of x, that is equivalent to the fact that $S(x, y)$ is TP_2 (DTP(1,1) dependence, see Chapter 3, Section 3.2.7).

(5) l and the upper Fréchet bound :
The density $f^+(u, v)$ of the upper Fréchet bound on the unit square can be viewed as the Dirac measure on the first diagonal $\delta_u(v)$ (see discussion of Fréchet bounds in Chapter 4).
What is the limit of l ? If this limit exists, it is :

$$l^+(u, v) = \frac{f^+}{S^+} - \frac{D_1 S^+ . D_2 S^+}{S^+ . S^+}$$

(the symbol + corresponds to the upper Fréchet bound).
We have

$$S = 1 + F - F_1 - F_2.$$

Therefore :

$$D_1 S = D_1 F - D_1 F_1.$$

Here $D_1 F^+(u, v) = 1_{[0,v]}(u)$, hence $D_1 S^+(u, v) = 1_{[0,v]}(u) - 1_{[0,1]}(u)$ i.e. $D_1 S^+(u, v) = -1_{]v,1[}(u)$.
In the same manner one can deduce that $D_2 S^+(u, v) = -1_{]u,1[}(v)$.
Therefore $D_1 S(u, v) D_2 S(u, v) = 0$ and $l(u, v) = \frac{\delta_u(v)}{S}$. The upper bound of l is thus concentrated on the diagonal of the unit square.

(6) A multivariate generalization [52] :

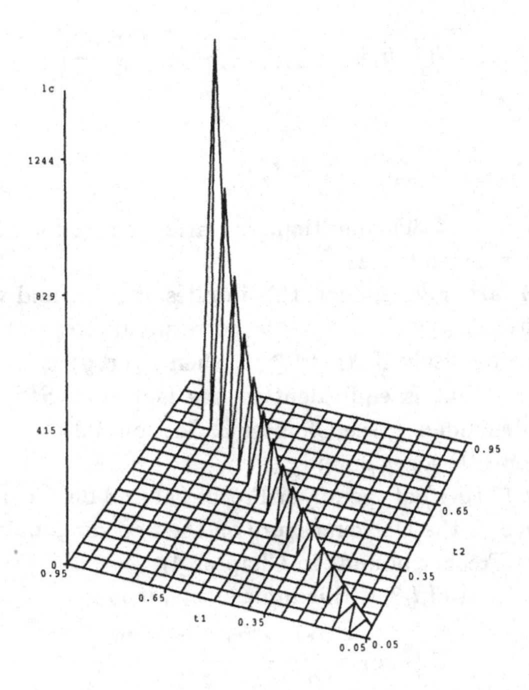

Fig. 6.4 l for Clayton's copula with high dependence, $\alpha = 200$.

Let S be the survival function of a non-negative random vector $\mathbf{X} = (X_1, X_2, ..., X_n)$. Then :

$$S = S_1 S_2, ..., M_{12} M_{13} ... M_{123} ... M_{12..n},$$

where S_i, $i = 1, ..., n$ are the univariate marginal survival functions of S and the interactions terms M_{ij} are defined from the bivariate marginals $S_{ij} = S_i S_j M_{ij}$; or also $M_{ij} = \exp \int_{[0,x_i]x[0,x_j]} dLogS_{ij}$, the last term being $M_{1,2,...,n} = \exp \int_{[0,\mathbf{x}]} dLogS$. The functions $D_i D_j LogS$,...,$dLogS$ are then multivariate generalizations of the conditional covariance rate $l(x, y)$. The increments of $LogS$ on the hyper-cube $[0, \mathbf{x}]$ can be interpreted in different ways. For example:

$$\int_{[0,\mathbf{x}]} dLogS = \Sigma_{|\alpha \geq 2|} (-1)^{\nu - \alpha} \left[LogS_\alpha(\mathbf{x}_\alpha) - Log(\Pi_{i \in \alpha} S_i(x_i)) \right]$$

where α denotes a set of indices: $\emptyset \neq \alpha \subseteq \nu = \{1, 2, ..., n\}$, and $\mathbf{x}_\alpha$ denotes the coordinates of the point $\mathbf{x}$ whose indices fall into the set α. The right-hand side can be interpreted as a weighted sum of dependence functions comparing each joint marginal S_α with its one-dimendional marginals.

Another form is :

$$\int_{[0,\mathbf{x}]} dLogS = -\Sigma_{(\tau_1,...,\tau_l)}(-1)^{l-1}(l-1)!\left[LogS(\mathbf{x}) - Log(\Pi_{i=1}^{l} S_{\tau_i}(\mathbf{x}_{\tau_i})\right]$$

where the sums is over the partitions $\tau = (\tau_1, ..., \tau_l)$, $l > 1$, of the set $\nu = \{1, 2, ..., n\}$ into non-empty mutually disjoint blocks. The right-hand side can be interpreted as a weighted sum of measures of dependence in which the term associated with the partition τ is a measure of deviation of the survival function S from $S_{H_0} = S_{\tau_1}.S_{\tau_2}...S_{\tau_l}$. This measure takes into account the dependence which is not contained in the lower dimensional marginals. In [52], Dabrowska studies the properties of these dependence functions and proposes a class of tests of independence for censored data.

6.3.8.3 Θ : *the ratio of two conditional hazards*

The index Θ is the ratio of two conditional hazards :

$$\Theta(x, y) = \frac{h_{2/1}(x, y)}{h_2(x, y)} \tag{6.16}$$

i.e., using (1.1)

$$\Theta(x, y) = \frac{D_2 D_1 S(x, y) S(x, y)}{D_1 S(x, y) D_2 S(x, y)} \,. \tag{6.17}$$

Alternatively :

$$\Theta(x, y) = \frac{h(x, y)}{h_1(x, y) h_2(x, y)} \,. \tag{6.18}$$

This index was originally defined and used by Clayton and Cuzick [46], Oakes [163] and Anderson *et al.* [6].

(1) Interpretation :
Θ was defined by Clayton and Cuzick [46] as an odds-ratio. Suppose we are dealing with two discrete failure time variables taking on the values $i = 1, 2, ..., n$ and $j = 1, 2, ..., m$. Denote by P_{ij}, the relative frequency of a failure of the two components in the cell (i, j), then the discrete analogue of Θ is:

$$\Theta(i, j) = \frac{P_{ij} \sum_{l>i} \sum_{k>j} P_{lk}}{\sum_{l>i} P_{lj} \sum_{k>j} P_{ik}} . \tag{6.19}$$

(2) Θ and the joint distribution :
- Θ is symmetric,
- Θ is marginal–free. Indeed, using the formulas shown in the preceding section, $D_1 D_2 S = c(S_1, S_2) f_1 f_2$ and $D_1 S = D_1 C(S_1, S_2) f_1$ and similarly for $D_2 S$, we obtain :
$\Theta = \frac{C D_2 D_1 C}{D_1 C D_2 C}$ where C is the survival copula associated with S.
- Θ and the two conditional hazards h_1 and h_2 determine the joint distribution : Indeed, using equations (6.18) and (1.2), we arrive at

$$\Theta - 1 = \frac{l}{h_1 h_2}.$$

We know that l together with S_1 and S_2 determines S.

$$\frac{S}{S_1 S_2} = \int_0^x \int_0^y l(u, v) du dv. \tag{6.20}$$

However using $h_1 = -\frac{D_1 S}{S}$, we obtain :

$$\frac{S}{S_1} = exp(-\int_0^x h_1(u, y) du). \tag{6.21}$$

Thus the two preceding equations allow us to determine S_2. In the same manner, S_1 can be determined.

(3) Θ and dependence :
- If (X, Y) are independent, $\Theta \equiv 1$. Namely, in that case $h_{2/1} = h_2 = h_y$, the marginal hazard of Y. The converse is also true. The stronger the dependence between X and Y is, the higher is the value of Θ.
- Θ and the upper Fréchet bound :

Using the results presented in the preceding section dealing with Fréchet bound of l, we observe that Θ approaches infinity when S tends to its upper Fréchet bound.
- The condition $\Theta \geq 1$ is equivalent to S being TP_2: indeed, in this case $l \geq 0$ and therefore , using Eq.(1.3), $-D_1 h_2 > 0$, i.e. h_2 is a decreasing function of x, this is equivalent to S being TP_2 (Chapter 3, Section 3.2.7).
-Oakes has shown that $\frac{\Theta-1}{\Theta+1}$ is a conditional version of Kendall's τ. More precisely, using the same notations that in Section 6.2.6,

$$\frac{\Theta(x,y)-1}{\Theta(x,y)+1} = E(Z_{ii'}|min(X_i, X_{i'}), min(Y_i, Y_{i'})).$$

Therefore Θ also measures a "remaining" dependence, but this is not exactly the same as the measure obtained with $\tau(x_0, y_0)$ in Section 6.4.3.
- Θ being constant is equivalent to the condition

$$S(x,y)^{1-\Theta} = S_1^{1-\Theta} + S_2^{1-\Theta} - 1\,.$$

Proof : One can write Equation (6.17) in the form :

$$\frac{D_1 D_2 S}{D_1 S} = \Theta \frac{D_2 S}{S}\,.$$

Integrating with respect to y, we obtain :

$$\log(-D_1 S) = \Theta Log S + k(x)\,.$$

We shall evaluate the function $k(x)$, at the point $(x, 0)$:

$$\log(-D_1 S_1) = \Theta Log(S_1) + k(x)\,.$$

Substituting $k(x)$ with its explicit expression, we arrive at

$$\log(-D_1 S) = \Theta log(\frac{S}{S_1}) + log(-D_1 S_1)\,,$$

$$\log\left(\frac{D_1 S}{D_1 S_1}\right) = \Theta \log(\frac{S}{S_1}).$$

Exponentiating, we have :

$$\frac{D_1 S}{D_1 S_1} = \left(\frac{S}{S_1}\right)^{\Theta}\,,$$

i.e.

$$\frac{D_1 S}{S^{\Theta}} = \frac{D_1 S_1}{S_1^{\Theta}} ,$$

and integrating with respect to x, we obtain

$$\frac{S^{1-\Theta}}{1-\Theta} = \frac{S_1^{1-\Theta}}{1-\Theta} + j(y)$$

for some $j(y)$. Using the symmetry of Θ, one can also write

$$\frac{S^{1-\Theta}}{1-\Theta} = \frac{S_2^{1-\Theta}}{1-\Theta} + m(x)$$

for some $m(x)$. Thus :

$$S^{1-\Theta} = S_1^{1-\Theta} + S_2^{1-\Theta} + const\,.$$

Evaluating the constant at the point (0,0), we have its value to be -1, hence we thus obtained the required formula.
- In the Clayton family, Θ is equal to a constant greater than one, but in many other families with positive dependence, $\Theta(x, y)$ is a decreasing function of x or y. This situation seems to be more realistic. Indeed, for small value of x, the event $\{X = x\}$ is more informative as far as the variable Y is concerning that the vague event $\{X > x\}$. In that case, h_2 is almost equal to h_y, the marginal hazard, but $h_{2/1}$ is probably very large, and therefore $\Theta(x, y)$ is large. As x increases the information provided by the event $(X > x)$ and $(X = x)$ becomes equivalent and hence Θ decreases to one.
For example, let $X = A + B$ and $Y = A + C$, with A, B, C being independent, with the same exponential standard exponential distribution $\Gamma(1, 1)$. In that case :

$$\Theta = \frac{|y-x| + 2 - exp(-min(x, y))}{|y-x| + 1 - exp(-min(x, y))}\,.$$

As x approaches infinity, Θ decreases from $\frac{|y|+1}{|y|}$ to 1.
- If Θ decreases to one when x or y tend to $+\infty$, then the dependence of the pair (X, Y) is at least $DTP(0, 1)$ or $DTP(1, 0)$, i.e. $h_{2/1}(x, y)$ and $h_{1/2}(x, y)$ decrease in x (respectively in y). Indeed: since Θ is greater than 1, $h_2(x, y)$ is decreasing with x, and so is

$h_{2/1}(x, y)$, therefore the dependence is at least DTP(0,1). Since Θ is symmetric in x and y, the dependence is also DTP(1,0).
-Oakes [163], utilizing Eq. (6.17) and censored data if available, proposes an estimator of Θ when the data are discretized.

(4) A generalization : A multivariate definition of Θ.
Starting from Equation (6.17), one can straightforwardly generalize the definition of Θ in case of more than two variables by defining :

$$\Theta(\mathbf{x}) = \frac{S(\mathbf{x})D_k D_j S(\mathbf{x})}{D_k S(\mathbf{t}) D_j S(\mathbf{x})}$$

where $\mathbf{x} = (x_1, x_2, ..., x_m)$ and $D_j = \frac{\partial}{\partial x_j}$. As in the case of two variables, it is the ratio of two hazards . If, for example $\mathbf{X} = (X_1, X_2, ..., X_m)$ is the random vector associated with the failures of members of a cluster or a family, then $\Theta(\mathbf{x})$ is the ratio of $-\frac{D_k D_j S(\mathbf{x})}{D_k S(\mathbf{x})}$, the hazard of the member j given that the member k failed at $X_k = x_k$ while the other members survived beyond x_i for $i = 1, 2, ..., j-1, j+1, ..., k-1, k+1, ..., n$, and $-\frac{D_j S(\mathbf{x})}{S(\mathbf{x})}$, the hazard of member j given $X_i > x_i$ for $i = 1, 2, ..., j-1, j+1, ..., n$. In the case of bivariate frailty models Oakes [163] has shown (see below Section 6.5.2) that $\Theta(\mathbf{x})$ can be written as a function of the survival function S only : $\Theta(\mathbf{x})) = \Theta^*(S(\mathbf{x}))$. In that case, Cho Paik *et al.* [45] have extended this definition to the case of multivariate frailty models allowing to take into account the failures of two (or more) members of the cluster in the hazard of a member. For example :

$$\Theta^*(D_m S(\mathbf{x})) = \frac{D_m S(\mathbf{x}) D_k D_j D_m S(\mathbf{x})}{D_k D_m S(\mathbf{x}) D_j D_m S(\mathbf{x})}$$

is the ratio of two hazards: the hazard of the member j given that the member k and the member m have failed while the other members have not, and the hazard of the member j given that the member m have failed while the other members have not.
Cho Paik *et al.* [45] have also shown that in case of this model there are recursive relations between the functions $\Theta^*(S(\mathbf{x}))$, $\Theta^*(D_m S(\mathbf{x}))$, $\Theta^*(D_m D_k S(\mathbf{x}))$, etc. More exactly, if $u = \varphi(S(\mathbf{t})) =$

$\Sigma_j \varphi(S_j(t_j))$, then the ratio Θ can be written as :

$$\Theta(u) = \frac{\varphi(u)\varphi^{(2)}(u)}{(\varphi^{(1)}(u))^2},$$

where $\varphi^{(j)}(u)$ denotes the jth derivative of $\varphi(u)$, with respect to u. Also since $D_k D_j(u) = 0$, $\forall j$, $\forall k = 1, ..., n$, we have :

$$\Theta^*(D_m S(\mathbf{x})) = \Theta_{(1)}(u) = \frac{\varphi^{(1)}(u)\varphi^{(3)}(u)}{(\varphi^{(2)}(u))^2}.$$

There are recursive relationships among $\Theta_{(1)}(u)$,$\Theta_{(2)}(u)$, ...,$\Theta_{(n)}(u)$:

$$\varphi^{(n+1)}(u)\Theta_{(n+1)}(u) = 2\varphi^{(n+1)}(u) - \frac{d}{du}\frac{\varphi^{(n)}(u)}{\Theta_{(n)}(u)}.$$

That is a choice of the distribution of the frailty variable determines all the dependence functions between the members of the cluster.

6.3.8.4 *Other measures derived from* Θ

To provide good estimators of dependence measures in case of censored data, Fan *et al.* [67] propose to use weighted dependence measures independent of the censoring distribution. For example, instead of $\Theta(x,y) = \frac{h(x,y)}{h_1(x,y)h_2(x,y)}$, they propose to use the weighted dependence measure :

$$D(x,y) = \frac{\int_0^x \int_0^y h(u,v)dudv}{\int_0^x \int_0^y h_1(u,v)h_2(u,v)dudv} = \frac{H(x,y)}{\int_0^x \int_0^y h_1(u,v)h_2(u,v)dudv} \tag{6.22}$$

where the instantaneous hazards are replaced by the cumulative hazards.

More generally, the weighted dependence measures is of the form :

$$D(x,y) = \frac{\int_0^x \int_0^y \phi(\Theta(u,v))w(du,dv)}{\int_0^x \int_0^y w(du,dv)}$$

where ϕ is a fixed function (for example $\phi(c) = c$) and w is a weight function equal to

$$w = \frac{D_1 S D_2 S}{S}$$

in Eq. (6.22).

The weight function is chosen such that it does not depend on the censoring distribution. The ϕ function is chosen to avoid no integrable differential function in the denominator which leads to estimators defined only on the grid formed by the uncensored failure times.

Shih and Louis [207] have proposed other weighted measures based on the covariance process of the martingale residuals (see Section 6.3.8.1) for testing independence of bivariate survival data.

6.3.8.5 *A local measure of LRD dependence : γ_f*

(1) Definition of γ_f :
Holland and Wang [98] defined $\gamma_f = \frac{\partial^2 f}{\partial x \partial y} = \nabla logf$, where $f(.,.)$ is a bivariate density. They show that this index is a continuous version of $\alpha(i,j)$, the log of a cross-ratio between four adjacent cells in an ordinal contingency table, given by :

$$\alpha(i,j) = \log(\frac{f_{ij} f_{i+1,j+1}}{f_{i+1,j} f_{i,j+1}}).$$

Specifically :

$$\gamma_f(x,y) = \lim_{dx \to 0, dy \to 0} \frac{1}{dxdy} \ln \left(\frac{f(x,y) f(x+dx, y+dy)}{f(x+dx,y) f(x,y+dy)} \right). \quad (6.23)$$

(2) The joint distribution :
- γ_f is symmetric.
- γ_f is marginal free. Namely $\gamma_f = \gamma_c$, where c is the density of the associated copula.
- If (X,Y) has a bivariate normal distribution with the correlation coefficient ρ , then $\gamma_f = \frac{\rho}{1-\rho^2}$. In this case this index is a constant. Conversely if γ_f is a constant, it is straightforward to show that the density $f(x,y)$ is of the form

$$a(x;\theta)b(y;\theta)exp(\theta xy).$$

See Jones [119].

(3) Dependence :
Evidently, if the components of the pair (X,Y) are independent, $\gamma_f \equiv 0$.
Equation (6.23) enables us to deduce that the condition

$$\gamma_f(x,y) \geq 0, \quad \forall x, \forall y$$

is equivalent to $f(x,y)$ being TP_2. Hence $\gamma_f(x,y)$ is an appropriate index for measuring local LRD dependence.
Jones [118] has shown, using a kernel method, that $\gamma_f(x_0,y_0)$ is a local version of the linear correlation coefficient.

6.4 Non-parametric Estimation of Local Indices

To estimate l or Θ, we ought to estimate the bivariate hazard $h(x,y)$, and the two conditional hazards $h_1(x,y)$ and $h_2(x,y)$.

6.4.1 *The univariate case : estimation of $H(x)$*

We shall briefly survey estimation of $H(x)$ and proceed to the bivariate case at hand. In the univariate case, with a random sample $(X_1, X_2, ..., X_i, ..., X_n)$ we associate the ordered sample $(X_{(1)}, X_{(2)}, ..., X_{(i)}, ..., X_{(n)})$. An empirical estimate of $H(x)$ – the cumulative hazard – is the Nelson-Aalen [162] estimate :

$$\widehat{H}_n(x) = \sum_{x_{(i)} \leq x} \frac{1}{n-i+1}.$$

To obtain a smooth estimate of $h(x)$, one considers $h(x)$ as the derivative of $H(x)$ and the smoothing is obtained by means of a convolution-product with a kernel K. Specifically :

$$\widehat{h_n(x)} = \int K_{b_n}(x-u)dH_n(u) \tag{6.24}$$

$$= \frac{1}{nb_n}\sum_{i=1}^{n}\frac{1}{n-i+1}K(\frac{(x-X_{(i)})}{b_n}. \tag{6.25}$$

Here, b_n is the window width, K is a bounded mapping from R to R with

$$\int_{\infty}^{+\infty} K(x)dx = 1,$$

satisfying

$$\lim_{x\to\infty} |x|K(x) = 0.$$

To assure that the obtained estimate $h_n(x)$ is consistent, the window width has to satisfy,

$$\lim_{n\to\infty} b_n = 0$$

and

$$\lim_{n\to\infty} nb_n = \infty \, .$$

A measure of discrepancy between the estimator $\widehat{h_n(x)}$ from the actual hazard $h(x)$ is the mean square error at x :

$$MSE_x(\hat{h}) = E(h_n\hat{(}x) - h(x))^2 \, .$$

The right-hand side term can be decomposed into two terms : bias and variance .

$$MSE_x(\hat{h}) = [E\widehat{h_n(x)} - h(x)]^2 + var\widehat{h_n(x)}.$$

A kernel K which is asymptotically optimal is the Epanechnikov kernel

$$K(x) = \frac{3}{4\sqrt{5}}(1 - \frac{x^2}{5})1_{[-\sqrt{5},\sqrt{5}]}.$$

It assures that the mean integrated square error (MISE)

$$MISE = E(\int (\widehat{h(x)} - h(x))^2 dx)$$

will be minimum. In the univariate case the theoretical optimal choice for the window width, when the criterion is MISE, is $b_n \propto n^{-\frac{1}{5}}$.

6.4.2 *Bivariate case and conditional hazards*

The bivariate hazard $H(x, y)$ is estimated by an analogue of the Nelson-Aalen estimate [1], [162] :

$$\widehat{H_n}(x,y) = \sum_{(x_i,y_i)\,:x_i\leq x, y_i\leq y} \frac{1}{n-i+1} \, .$$

In this case $\widehat{h}_n(x,y)$ can also be obtained using a convolution-product by means of a bivariate kernel , $K_{b_{1n},b_{2n}}(u,v)$:

$$h_n\widehat{(x,y)} = \int\int K_{b_{1n},b_{2n}}(x-u, y-v)\widehat{H_n}(du,dv) \, .$$

The two conditional hazards $h_i(x,y)$, $i = 1,2$, are obtained in the same manner from the partial cumulative hazards. For example :

$$\widehat{H_1}(x,y) = \sum_{(x_i,y_i)\,:y_i \leq y, x \leq x_i < x+dx} \frac{1}{n-i+1}.$$

Hence

$$\widehat{h_1}(x,y) = \int_{[0,x]} K_{b_{1n}}(x-u)\widehat{H_1}(du,y).$$

6.4.2.1 *Consistency and asymptotic normality of the estimate of* $h(x,y)$

Dabrowska *et al.* [53] have established conditions to assure consistency and asymptotic normality of these estimators, in a general case where the data are censored. If the data are not censored, and if the following conditions are valid :

(1) compact support :
$(x,y) \in J = [s_1,s_2] \times [t_1,t_2] \subseteq J_\epsilon$ where $\epsilon > 0$ is small and $J_\epsilon = [s_1-\epsilon, s_2+\epsilon] \times [t_1-\epsilon, t_2+\epsilon] \subseteq [0,\tau]$ with $S(\tau) > 0$.
(2) conditions on $h(x,y)$:

 (a) $h(x,y)$ is bounded and strictly positive :

 $$0 < \inf\{h(x,y) \ : (x,y) \in J_\epsilon\} < \sup\{h(x,y) \ : (x,y) \in J_\epsilon\} <$$

 (b) $h(x,y)$ is twice continuously differentiable in (x,y).

(3) on the kernel :
- $K(x,y)$ is a function of bounded variation on $[-1,1]^2$, vanishing on the boundary of this set,
-

$$\int \mathbf{u}K(\mathbf{u})d\mathbf{u} = 0\,,$$

$$\int \mathbf{u}\mathbf{u}'K(\mathbf{u})d\mathbf{u} = d(K)I$$

where I is the identity matrix, and $d(K)$ is a finite constant, and $c(K) = \int K^2(\mathbf{u})d\mathbf{u}$.
(4) on the bandwidth : There exist sequences $b_{in}, i = 1,2$, which satisfy

(a) $max(b_{1n}, b_{2n}) \to 0,$
(b) $nb_{1n}b_{2n} \to \infty$
(c) $logn = o(nb_{1n}b_{2n})$

are valid. Then $\widehat{h_n}(x, y)$ is a consistent estimator of $h(x, y)$, asymptotically normal, with an asymptotic mean square error (AMSE) at almost all points (x, y) given by :

$$AMSE_{\hat{h}}(x,y) = \frac{c(K)h(x,y)}{nb_{1n}b_{2n}S(x,y)} + \frac{d(K)^2}{4}(\sum_{i=1}^{2} b_{in}^2 \nabla_i^2 h(x,y))^2 .$$

The first term on the right-hand side is the variance, the second is the bias. One observes that at the upper tail of the distribution, where $S(x, y)$ is small, the variance of the estimator is large.

6.4.2.2 *Consistency and asymptotic normality of the two conditional hazards*

Conditions for consistency and asymptotic normality of the conditional hazards are similar to those for the bivariate hazards. There are as follows:

(1) compact support :
Let $(x, y) \in J_{1\epsilon} = [s_1 - \epsilon, s_2 + \epsilon] \times [0, t_2]$, and $J_{2\epsilon} = [0, t_1] \times [t_1 - \epsilon, t_2 + \epsilon]$ where $J_{i\epsilon} \subseteq [0, \tau]$ with $S(\tau) > 0$;

(2) on $h_i(x, y)$:
  (a) $h_i(x, y)$, $i = 1, 2$ is bounded and strictly positive ;
  (b) $0 < \inf\{h_i(x, y) \ : (x, y) \in J_{i\epsilon}\} < \sup\{h(x, y) \ : (x, y) \in J_{i\epsilon}\} < \infty$;
  (c) $h_i(x, y)$ $i = 1, 2$ is twice continuously differentiable with respect to the ith coordinate of (x, y).

(3) on the kernel :
- $\bar{K}(x)$ is a function of a bounded variation on $[-1, 1]$, vanishing on the boundary of this set,
- $\int u\bar{K}(u)du = 0$, $\int u^2 K(u)du = d_1(\bar{K})$, a finite constant, and $c_1(\bar{K}) = \int \bar{K}^2(u)du$.

(4) on the bandwidth :
There exist sequences a_{in}, $i = 1, 2$ which satisfy :
- $max(a_{in}) \to 0$, $na_{in} \to \infty$
- $logn = o(na_{in})$.

If these conditions are fulfilled, then for almost all (x,y) in $J_{i\epsilon}$, $h_{in}\widehat{(x,y)}$ $i = 1,2$, is a consistent estimator of $h_i(x,y)$, asymptotically normal, with an asymptotic mean square error (AMSE) at the point (x,y) given by :

$$AMSE_{\hat{h}_i}(x,y) = \frac{c_1(\bar{K})h_i(x,y)}{na_{in}S(x,y)} + \frac{d_1(\bar{K})^2}{4}[\nabla_i^2 h_i(x,y)]^2,\ i=1,2.$$

6.4.3 *Estimation of the indices l and Θ*

Having estimators for $h(x,y)$, $h_1(x,y)$ and $h_2(x,y)$, we can obtain estimators for $l(x,y)$ and $\Theta(x,y)$. The consistency and asymptotic normality of $l(x,y)$ and $\Theta(x,y)$ are obtained using the delta method . If we choose an optimal bandwith for $h_1(x,y)$ and $h_2(x,y)$, then the rate of convergence for $\widehat{h_1}(x,y)$, and $\widehat{h_2}(x,y)$ will be $n^{-4/5}$, which is faster than the rate for $\hat{h}(x,y)$, which is $n^{-2/3}$. Namely only the variance of $\hat{h}(x,y)$ contributes to the variance of $\hat{l}(x,y)$ and $\hat{\Theta}(x,y)$. In other words

$$varl\widehat{(x,y)} \approx varh\widehat{(x,y)}.$$

6.5 A Search for the Localisation of the Maximal Association

We shall illustrate this search by means of three families of Archimedean copulas previously studied in the chapter on copulas. Here, we shall use the definition from the survival functions.

(1) The Clayton family : This family verifies $\Theta = const.$ As we shall see that yields a "late" dependence :

$$S(x,y) = ((S_1(x))^{1-\alpha} + (S_2(y))^{1-\alpha})^{1/\alpha-1},\ \alpha > 1.$$

(2) The Gumbel-Hougaard family :

$$S(x,y) = \exp\left[-\left((-\ln(S_1(x)))^{1/\alpha} + (-\ln(S_2(y)))^{1/\alpha}\right)^{\alpha}\right],\ 0 < \alpha <$$

In this case $\Theta(s) = 1 + (1-\alpha)/(-\alpha\ln(s))$ and Θ decreases from $+\infty$ to 1 when S decreases from 1 to 0. This property characterizes an "early" dependence.

(3) The Frank family :

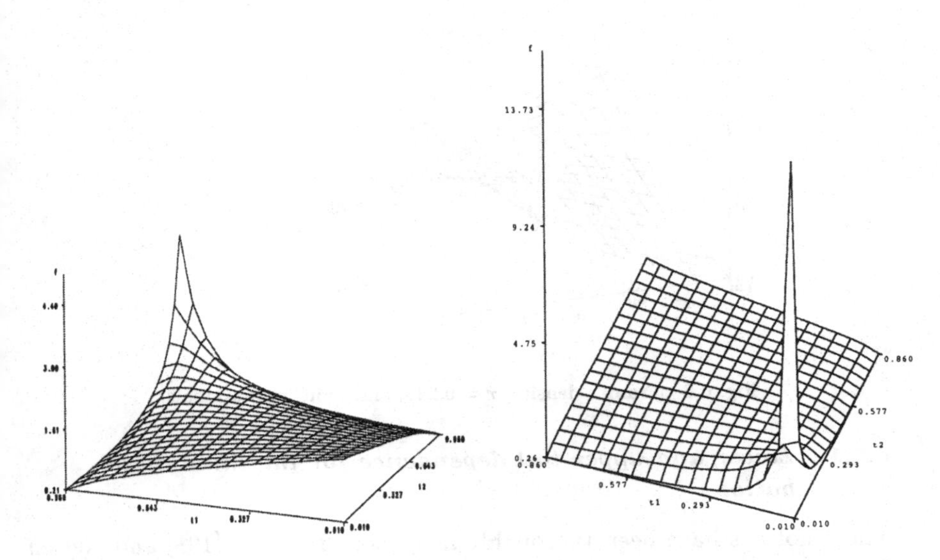

Fig. 6.5 Densities with uniform marginals and τ = 0.243, a) Clayton's model b) Gumbel-Hougaard model.

$$S(x,y) = \log_\alpha \left(1 + \frac{(\alpha^{S_1 - 1})(\alpha^{S_2 - 1}))}{(\alpha - 1)}\right) \text{ ,} 0 < \alpha < 1.$$

In this case the index $\Theta(s) = -s\frac{ln(\alpha)}{(1-\alpha^s)}$ is decreasing from $ln(\gamma)/(\gamma - 1)$ to 1 when S decreases from 1 to 0. Here the dependence is "median".

We represent the three families for the same τ, i.e. the same global dependence. Note the peak in the case of Clayton model near the point (1,1), while for the Gumbel-Hougaard model it is at (0,0). Note also a saddle in the vicinity of (1/2,1/2) for Frank's model.

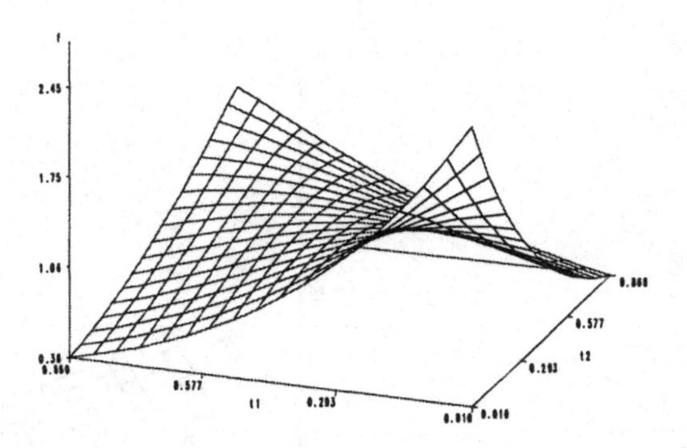

Fig. 6.6 Frank's density, $\tau = 0.243$, and uniform marginals.

6.5.1 *Lower and upper tail dependence for the three distributions*

These notions have been thoroughly investigated by Joe [108] and Nelsen [160]. Above we have defined these three copulas by means of their survival functions S, S_1 and S_2. Therefore employing the notations of Section 4.2 of Chapter 4, we shall utilize copulas $C'(u,v)$, recalling that the upper tail dependence of $C'(u,v)$ is the lower tail dependence of $C(u,v)$ and vice versa. For example, if we are searching for the upper tail dependence then:

$$\delta = lim_{u\to 1}\frac{\bar{C}'(u,u)}{1-u} = lim_{u\to 0}\frac{C(u,u)}{u} \ .$$

However $C(u,u) = \varphi^{-1}(2\varphi(u))$, therefore :

$$\delta = lim_{u\to 0}\frac{\varphi^{-1}(2\varphi(u))}{u} = lim_{t\to\infty}\frac{\varphi^{-1}(2t)}{\varphi(t)} \ .$$

For the Clayton family $\varphi^{-1}(t) = (t+1)^{\frac{-1}{\alpha-1}}$. Hence :

$$\delta = lim_{t\to\infty}\frac{(2t+1)^{\frac{-1}{\alpha-1}}}{(t+1)^{\frac{-1}{\alpha-1}}} \ .$$

Consequently

$$\delta = 2^{\frac{-1}{\alpha-1}} \ .$$

The Clayton family therefore possesses an upper tail dependence, which we shall call late dependence.

For the Gumbel-Hougaard and Frank families , we find $\delta = 0$. In these cases the dependence is not late.

Computing the lower tail dependence we obtain :

$$\gamma = lim_{u\to 0}\frac{C'(u,u)}{u} = lim_{u\to 1}\frac{\bar{C}(u,u)}{1-u}.$$

However $C(\bar{u}, u) = 1 - 2u + C(u,u) = 1 - 2u + \varphi^{-1}(2\varphi(u))$. Therefore

$$\gamma = lim_{u\to 1} 2 - \frac{1-\varphi^{-1}(2\varphi(u))}{1-u} = 2 - lim_{t\to 0}\frac{1-\varphi^{-1}(2t)}{1-\varphi^{-1}(t)}.$$

For the Gumbel-Hougaard family $\varphi^{-1}(t) = exp(-t^{\alpha})$, we obtain $\gamma = 2 - 2^{\alpha}$. The Gumbel-Hougaard family possesses a lower tail dependence, i.e. an "early" dependence. For the Clayton and Frank families these limits are zero.

6.5.2 Θ *and the remaining dependence*

Oakes [163] has shown that for Archimedean copulas, Θ is a function of S, and that this property is characteristic. Specifically :

$$\Theta(x,y) = \Theta^{*}(S(x,y))$$

where

$$\Theta^{*}(s) = \frac{-s\varphi''(s)}{\varphi'(s)}$$

$\varphi'(s)$ and $\varphi''(s)$ being the first and second derivative of φ. We can, therefore, easily represent the "evolution" of Θ for the three families discussed above.

As it was indicated above for the Gumbel-Hougaard family, $\Theta(s) = 1 + \frac{1-\alpha}{-lns}$, which decreases from infinity to 1 as S decreases from 1 to zero. The large values of Θ when S is in the vicinity of one, correspond to an "early" dependence for this family. We observe from Figure 6.7 that Θ is the largest for the Hougaard family when $S > 0.75$.

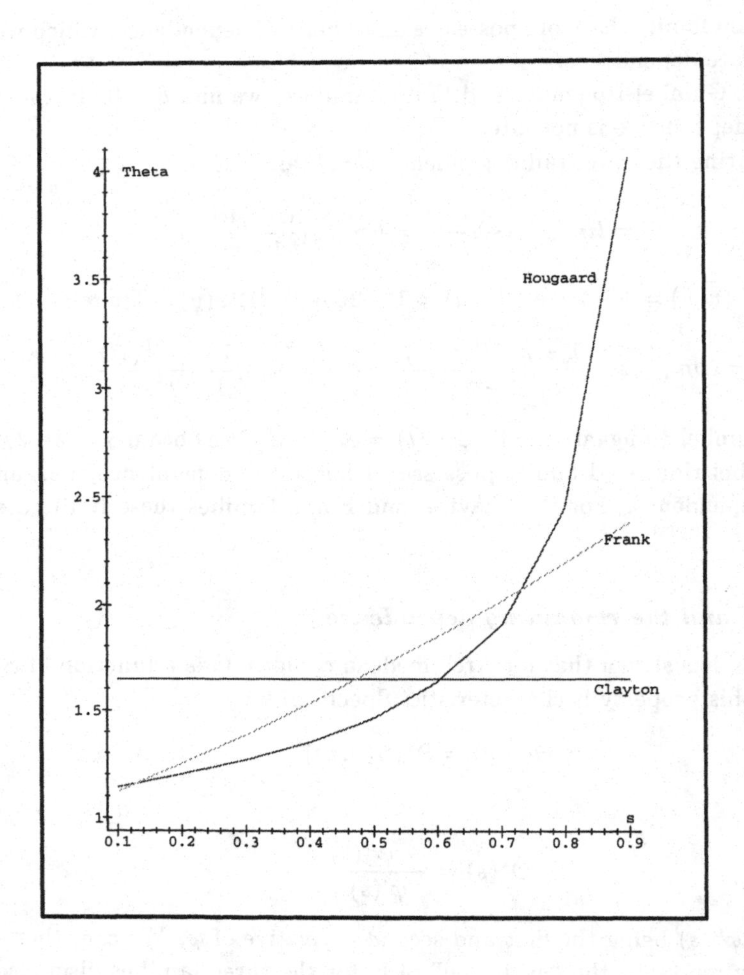

Fig. 6.7 Evolution of Θ as a function of S for the Clayton, Hougaard and Frank families ($\tau = 0.243$).

For the Frank family, Θ decreases from a finite value $\frac{ln(\alpha)}{\alpha-1}$ to 1 when S decreases from 1 to zero. For this family we notice that Θ is the largest for the intermediate values of S (S between 0.7 and 0.5). In this family the dependence is therefore "median".

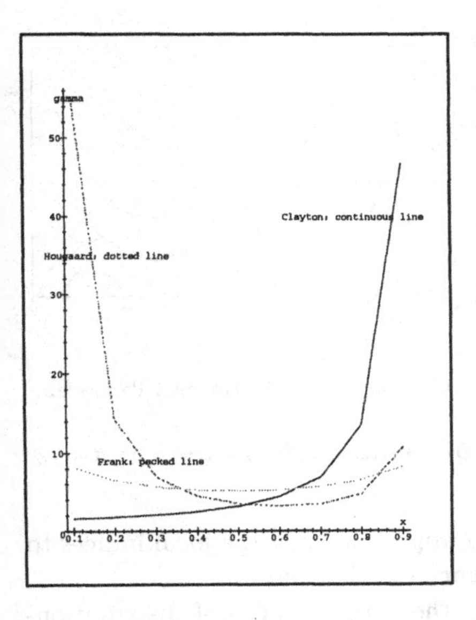

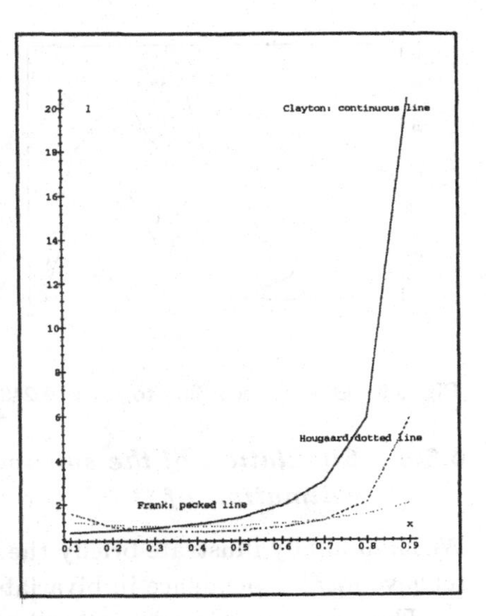

Fig. 6.8 Evolution of the indices $\gamma(x, x)$ and $l(x, x)$ as a function of x for the 3 families with $\tau = 0.243$.

For the Clayton family, Θ is constant, and hence Θ is the highest among the three distributions when S is small. From the figure 6.7 the corresponding values of S are from 0.5 to 0 . The dependence is "late".

Oakes [163] has shown that $\frac{\Theta-1}{\Theta+1}$ can be interpreted as a conditional τ, and in particular for the Clayton family the ratio of the concordant and discordant pairs is constant regardless of the value of S. This, of course, applies also to "late" dependence.

6.5.3 *Index γ and the instantaneous dependence*

As it was shown in Section 6.3.8.5, the index γ_f measures local LRD dependence . We describe this property for the three distributions. For the Clayton family, $\gamma_f(x, x)$ is maximal for x near to 1, conversely $\gamma_f(x, x)$ is maximal for x near to zero for the Hougaard family and the Frank family occupies intermediate position.

Note that the conditional covariance rate does not indicate in a simple manner the localization of the dependence.

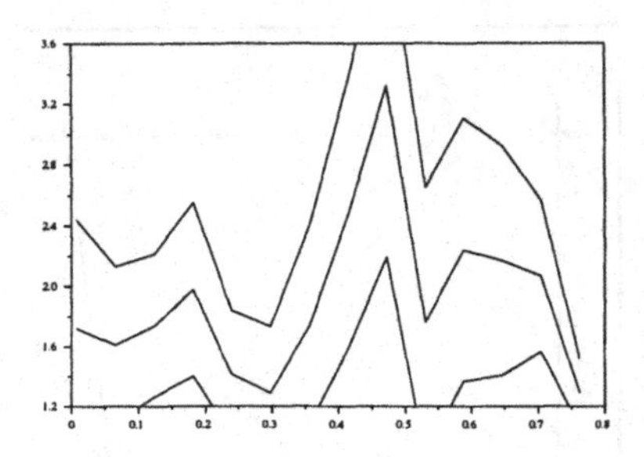

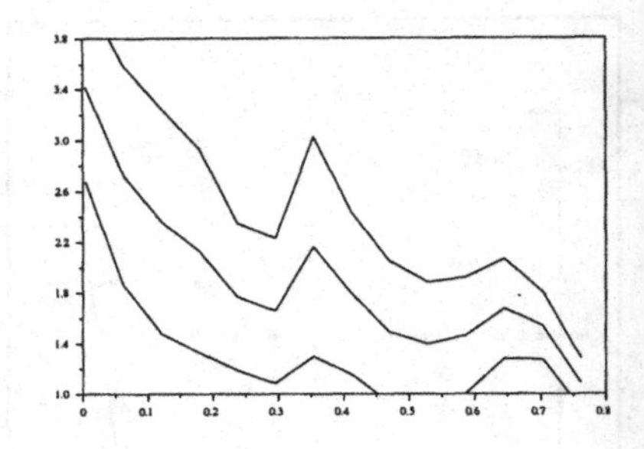

Fig. 6.9 $\Theta(x,x)$ according to x $\tau = 0.243$, a) Clayton's model b) Hougaard's model.

6.5.4 *Simulation of the survival bivariate distributions and estimation of Θ*

We shall finally illustrate briefly the appropriateness of the local indices to retrieve local dependence in bivariate survival distribution.

This is accomplished by simulating the three families of distribution-s Clayton's, Hougaard's and Frank's which served as models of different modes of dependence. The reader is referred to Section 4.10 in the chapter on Copulas for some details. We have used the method described by Genest [81] to simulate the distributions of Clayton's and Frank's families and the method described by Lee [141] to simulate the Hougaard family.

We shall use the Dabrowska's method (Section 5) to estimate Θ. Figures 6.9 and 6.10 provide Θ and the 0.90-confidence limit obtained by the jacknife method on the diagonal $y = x$. Variations of Θ as a function of x allow us to retrieve the shape of the dependence.

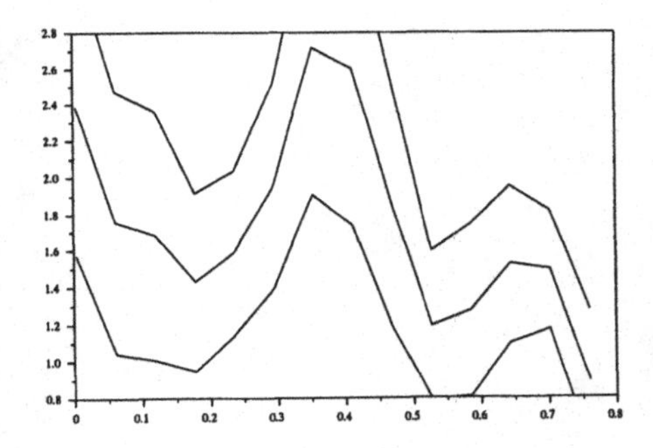

Fig. 6.10 $\Theta(x, x)$ as a function of x, $\tau = 0.243$, Frank's family.

Bibliography

[1] AALEN O., *Non parametric estimation of partial transition probabilities in multiple decrements models*, Ann. Statist, **6**, 534-545, 1978.

[2] ABEL N.H., *Recherche des fonctions de deux quantités variables indépendantes x et y ,telles que f(x,y), qui ont la proprieté que f(z,f(x,y)) est une fonction symétrique de z, x et y*, J. Reine Angew. Math., **1**, 11-15, Oeuvres complètes de N. H. Abel, 1826.

[3] ADRAIN R., *Research concerning the probabilities of the errors which happen in making observations*, The Analyst (or Mathematical Museum), 1, 93-109, 1807 .

[4] ALI M.M., MIKHAIL N.N., HAQ M.S., *A class of bivariate distributions including the bivariate logistic*, J. Multi. Anal, **8**, 405-412, 1978.

[5] AMINI M., AZARNOOSH H.A., BOZORGMA A., *The almost sur convergence of weighted sums of negatively random variables*, J. Sci. Isl. Rep. Iran, **10**, 112-116, 1999.

[6] ANDERSON J.E., LOUIS T.A., HOLM N.V., HARVALD B., *Time-dependent association measures for bivariate survival distributions*, J. Amer. Stat. Assoc.,**Vol 87** , 641-650, 1992.

[7] ANDERSON T.W., *An Introduction to Multivariate Statistical Analysis*, 2nd Edition, Wiley, New York, 1984.

[8] BAIRAMOV I.G., KOTZ S., *Dependence structure and symmetry of Huang-Kotz FGM distributions and their extensions*, Technical Report, The George Washington University, 2000a.

[9] BAIRAMOV I.G., BEKCI M., *Concomitant of order statistics in FGM type bivariate uniform distributions*, Journal of the Turkish Statistical Association, **2**, 135-144, 1999.

[10] BAIRAMOV I.G., KOTZ S., BEKCI M., *New generalized Farlie-Gumbel-Morgenstern distributions and concomitants of order statistics* Technical Report, The George Washington University, 2000.

[11] BAIRAMOV I.G., KOTZ S., *On a new family of positive quadrant*

dependent bivariate distributions, Technical Report, The George Washington University, 2000b.

[12] BAIRAMOV I.G., ERYILMAZ S.N., *The distribution of exceedance statistics in FGM sequences*, Unpublished manuscript, Ankara University, 2000.

[13] BAIRAMOV I., KOTZ S., KOZUBOWSKI T.J., *A new measure of local dependence*, Technical Report 362, University of California at Santa Barbara, 2000.

[14] BALASUBRAMANIAN K., BEG M.I., *Concomitant of order statistics in Morgenstern type bivariate exponential distributions* , J. Appl. Statist. , **5**, 233-245, 1997.

[15] BALASUBRAMANIAN K., BEG M. I. , *Concomitant of order statistics in Gumbel's bivariate exponential distribution*, Sankhya, **60**, Series B, 399-406, 1998.

[16] BARLOW R., PROSCHAN F., *Statistical theory of reliability and life testing*, Holt, Rinehart and Winston, New-York, 1975.

[17] BHATTACHARYA P.K. , *Induced order statistics: Theory and applications* In Handbook of Statistics, (P.R. Krishnaiah and P.K. Sen eds.), **4**, 383-403, North-Holland, Amsterdam, 1984.

[18] BÄUERLE N., *Inequalities for stochastic models via supermodular orderings*, Commun. Statist.-Stochastic Models, 13(1), 181-201, 1997.

[19] BAYES T., *An essay towards solving a problem in the doctrine of chance* , Phil. Trans. Roy. Soc., London, **53**, 370-418, 1763, [Reprinted in : Studies in the history of Statistics and Probabilities, London : Griffin].

[20] BELL C.B., *Mutual information and maximal correlation as measures of dependence*, 2, 587-595, 1961.

[21] BJERVE S., DOKSUM K., *Correlation Curves : Measures of association as function of covariate values*, Ann. Stat., **21**, Num 2, 890-902, 1993.

[22] BLOCK H.W., SAVITS T.H., SHAKED M., *Some concepts of negative dependence* , Ann. of Prob., **10**, 765-772, 1982.

[23] BLOCK H.W., GRIFFTHS W.S., SAVITS T.H., *L-superadditive structure functions*, Adv. Appl. Prob., **21**, 919-929, 1989.

[24] BLOCK H.W., SAMPSON A.R., *Conditionnally ordered distributions*, J. Multi. Anal., **27**, 91-104, 1988.

[25] BLYTH S.J., *Measuring local association : an introduction to the correlation curve*, Sociological Methodology, **24**, 171-197, 1994.

[26] BLYTH S.J., *Karl Pearson and the correlation curve*, Int. Stat. Rev., **62**, 3, 393-403, 1994.

[27] BORKOWF C.B., GAIL M.H., CARROL R.J., GILL R.D., *Analyzing bivariate continuous data grouped into categories defined by empirical quantities of marginal distributions*, Biometrics, **53**, 1054-1069, 1997 .

[28] BOWDITCH H.P., *The growth of children*, (Report of the Board of Health of Massachusetts VIII, 1877) [Reprinted in Papers in Anthropometry, Boston: 1894].

[29] BOWLEY A.L., *Measurement of Groups and Series*, London, 1903.

[30] BOWLEY A.L., *Elements of Statistics*, (3rd Edn.), New-York : Scribner 1907.

[31] BRADY B. SINGPURWALLA D., *Stochastically monotone dependence*, Technical report, Georges Washington University, 1996.

[32] BRAVAIS A., *Analyse mathémathique sur les probabilités des erreurs de situation d'un point*, Mem. Acad. Sci. Paris, Ser. 2, 255-332, 1846.

[33] BROWN W., *The Essentials of Mental Measurement*, Cambribdge, Cambridge University Press, 1911.

[34] BUTKIEWICZ J., HUS E., *On a class of bi-and multivariate distributions generated by marginal Weibull distributions*, Transactions of the 7th Prague conference on Information theory, Reidel, Dordrecht, Holland, 1977.

[35] CAMBANIS S., *Some properties and generalizations of multivariate Eyraud-Gumbel-Morgenstern distributions*, J. Multi. Anal., **7**, 551-559, 1977.

[36] CAMBANIS S., SIMONS G., STOUT W., *Inequalities for Ek(X,Y) when the marginals are fixed* , Z. Wahrscheinlichkeitstheorie verw. Gebiete, 36, 285-294, 1976.

[37] CAPERAÀ P., GENEST C., *Concepts de dépendance et ordres stochastiques pour des lois bidimensionelles*, The Canadian Journal of Statistics, **18**, 315-326, 1990.

[38] CAPÉRAÀ P., FOUGÈRES A.L., GENEST C., *Bivariate distributions with given extreme value attractor* , J. Multi. Anal., **72**, 30-49, 2000.

[39] CARLEY H., TAYLOR M.D., *A new proof of Sklar's theorem*, Conference on Distributions with given marginals, Barcelona, July 2000.

[40] CHAKAK A., KOEHLER K.J., *A strategy for constructing multivariate distributions*, Commun. Statist.-Simula., **24** (3), 537-550, 1995.

[41] CHANG C.-S., *A new ordering for stochastic majorization : Theory and applications* , Adv. Appl. Prob., **24**, 604-634, 1992.

[42] CHEYSSON E., *Review of Galton's "Regression towards mediocrity"* , in Le Temps, Oct. 23, 1885.

[43] CHHETRY D., KIMELDORF G., SAMPSON A.R., *Concepts of setwise dependence*, Prob. Eng. Inform Sci., 3, 367-380, 1989.

[44] CHHETRY D., KIMELDORF G., ZAHEDI H., *Dependence structures in which uncorrelatedness implies independence*, Stat. and Prob. Lett., 4, 197-201, 1986.

[45] CHO PAIK M., TSAI W.-Y., OTTMAN R., *Multivariate survival analysis using piecewise gamma frailty*, Biometrics, 50, 975-988, 1994.

[46] CLAYTON D., CUZICK J., *Multivariate generalizations of the proportional hazards model*, J.R. Statist. Soc. A, **148**, 2, 82-117, 1985.

[47] CLEMEN R.T., FISCHER G.W., WINKLER R.L., *Assessing dependence : some experimental results*, Management Science, **46**, 1100-115, 2000.

[48] COOK R.D., JOHNSON M.E., *A family of distributions for modeling*

non-elliptically symmetric data, J. Roy. Statist. Soc. Ser. B, **43**, 210-243, 1981.

[49] COHEN A., SACKROWITZ H.B., SAMUEL-CAHN E., *Cone order association*, J. Multi. Anal., **57**, 320-330, 1995.

[50] COLES S., HEFFERNAN J.,TAWN J., *Dependence measures for extreme value analyses*, Technical Report LA1 4YF, Lancaster University, February 2000.

[51] CURRIE J.E., Directory of coefficients of tail dependence, Dept of Math. and Statist., Techn. Report ST-99-06, Lancaster University, U.K., 1999.

[52] DABROWSKA D.M., *Weak convergence of a product integral dependence measure*, Scandinavian J. of Statist., 23, 551-580, 1996.

[53] DABROWSKA D.M., DUFY D.L., ZHANG D.Z., *Hazard and density estimation from bivariate censored data*, Nonparametric Statist., 1997.

[54] DAVENPORT C.B., *Statistical Methods with Special Reference to Biological Variation*, New-York Wiley, 1914.

[55] DAVENPORT C.B., *The Principles of Breeding*, New-York, 1907.

[56] DAVID H.A., *Order Statistics.* 2nd edn., John Wiley, New York, 1981.

[57] DAVID H.A., *Concomitant of order statistics*: Review and recent developments. In Multiple Comparisons, Selection and Application in Biometry (F. M. Hoppe, ed.), 507-518, Dekker, New-York, 1993.

[58] DE LA HORRA J., RUIZ-RIVAS C., *A Bayesian method for inferring the degree of dependence for a positively dependent distribution*, Commun. in Statist. Theor. Meth., **A17**, 4357-70, 1988.

[59] DE LA HORRAD J., FERNANDEZ C., *Sensitivity to prior independence via Farlie-Gumbel-Morgenstern model*, Commun. Statist. Theor. Meth., **24** (4), 987-996, 1995.

[60] DESROSIÈRES A., *The Politics of Large Numbers*, Cambribdge Mass. : Harvard University Press, 1998. (Translated from French).

[61] DOKSUM K., BLYTH S., BRADLOW E., MENG X., ZHAO H., *Correlation curves : Measures of association as local measures of variance explained by regression*, J. Amer. Stat. Assoc. , **89**, 571-582, 1994.

[62] DROUET MARI D., *La dépendance entre deux variables de durées : concepts de dépendance et mesures locales de liaison*, Rev. Stat. Appl., **XLV11**, 5-24 , 1999.

[63] EDGEWORTH F.Y., *Statistical correlation between social phenomena*, J. R. Statist. Soc, **56**, 670-675, 1893.

[64] ESARY J.D., PROSCHAN F., *Relationships among some concepts of bivariate dependence* , Ann. Math. Stat., **Vol 43**, 651-665, 1972.

[65] ESARY J.D., PROSCHAN F., WALKUP D.W., *Association of random variables with applications*, Ann. Math. Stat., **Vol 38**, 1466-1474, 1967.

[66] EYRAUD H., *Les principes de la mesure des corrélations*, Ann. Univ. Lyon, Sect. **A1**, 30-47, 1936.

[67] FAN J., PRENTICE R. L., HSU L., *A class of weighted dependence measures for bivariate failure time data*, J. Roy. Stat. Soc., Series B, **62**,

181-190, 2000.

[68] FANCHER R.E., *Galton on examinations : An unpublished step in the invention of correlation*, ISIS, **80**, 446-455, 1989.

[69] FANG Z., JOE H., *Further developments on some dependence ordering for continuous bivariate distributions*, Ann. Inst. Stat. Math., **44**, 501-517, 1992.

[70] FARLIE D.J., *The performance of some correlation coefficients for a general bivariate distribution*, Biometrika, **47**, 307-323, 1960.

[71] FECHNER G.T., *Kollektivmasslehre*, G.F. Lipps Ed., Engelmann, Leipzig, 1897.

[72] FRANK M.J., *On the simultaneous associativity of F(x,y)and x+y-F(x,y)*, Aequationes Math., **19**, 194-226, 1979.

[73] FRÉCHET M., *Sur les tableaux de corrélation dont les marges sont données*, Annales de l'Université de Lyon, Sec A, Ser. 3, **14**, 53-77, 1951.

[74] FRÉCHET M., Math. Magazine, 32, 265-288, 1958-1959.

[75] FREES E.W., VALDEZ E.A., *Understanding relationship using copulas*, North-American Actuarial Journal, **2**, 1, 1998.

[76] GALTON F., *Correlation and their measurement, chiefly from anthropometric data*, Proc. R. Soc London, **45**, 136-145, 1880.

[77] GALTON F., *Regression towards mediocrity in hereditary stature*, J. Anthropol. Inst., **15**, 246-260, 1885.

[78] GALTON F. , *Natural Inheritance*, London : MacMillan, 1889.

[79] GASSER T., MÜLLER H.G., MAMMITZSCH V., *Kernels for nonparametric curve estimation*, J. Roy. Statist. Soc., Series B, **47**, 238-252, 1985.

[80] GAUSS C.F., *Theoria Combinationis Observationum Erroribus Minimis Obnoxiae*, Göttingen, 1823.

[81] GENEST C., MACKAY R.J., *Copules archimédiennes et familles de lois bidimensionnelles dont les marges sont données*, Canadian J. Stat., **14**, 2, 145-159, 1986.

[82] GENEST C., *Frank's family of bivariate distributions*, Biometrika, **74**, 3, 549-55, 1987.

[83] GENEST C., RIVEST L.P., *Statistical inference procedures for bivariate Archimedean copulas*, J. Amer. Stat. Assoc., 423, **88**, 1034-1043, 1993.

[84] GENEST C., QUESADA MOLINA J.J., RODRÍGUEZ LALLENA J.A., *De l'impossibilité de construire des lois à marges multidimensionnelles données à partir de copules*, C.R. Acad Sci Paris, **320**, Série 1, 723-726, 1995.

[85] GHOUDI K. C., RIVEST L.P., *Propriétés statistiques des copules de valeurs extrêmes bidimensionnelles* , Canadian J. Stat., **26**, 187-197, 1998.

[86] GINI C., *Atti. R. Ist.*, lett. Arti., **74**, 185-213, 1914.

[87] GREENLAND S., *A lower bound for the correlation of exponentiated bivariate normal pairs*, Amer. Statist., **50** 2, 163-164, 1996.

[88] GUMBEL E.J., *Bivariate exponential distributions*, J. Amer. Stat. Assoc.,

55, 698-707, 1960.

[89] HACKING I., *The Emergence of Probability*, Cambridge : Cambridge University Press, 1975.

[90] HALD A., *History of Mathematical Statistics 1800-1930*, New-York Wiley, 1994.

[91] HALLIN M., SEOH M., *Is 131,000 a large sample size? A numerical study of Edgeworth expansions*, In Research developments in Probability and statistics, E. Brunner and M. Denker, Eds, VSP, utrecht, 141-161, 1998.

[92] HARRIS R., *A lower bound for the critical probability in a certain percolation models*, Proc. Camb. Phil. Soc., **56** , 13-20, 1960.

[93] HARRIS R., *A multivariate definition for increasing hazard rate distribution functions*, Ann. Math. Stat., **41**, 713-717, 1970.

[94] HASHORVA E., HÜSLER J., *Extreme values in FGM sequences*, J. Multi. Anal., **68**, 212-225, 1999.

[95] HOOKER R.H., *On the correlation of the marriage rate with trade*, J. Roy. Statist. Soc, **64**, 486, 1901.

[96] HOËFFDING W., *Collected Works*, in N.I. Fisher and P.K. Sen Eds, Springer-Verlag, 1994.

[97] HOËFFDING W., *A non-parametric test of independence* , Ann. Math. Statist, **19**, 546-557, 1948.

[98] HOLLAND P.W., WANG Y.J., *Dependence functions for continuous bivariate densities*, Commun. Statist.-Theory Meth., **16** (3), 863-876, 1987.

[99] HOUGAARD P., *A class of multivariate failure time distributions*, Biometrika, **73**, 671-678, 1986.

[100] HOUGAARD P., *Analysis of Multivariate Survival Data*, Springer-Verlag, New-York, 2000.

[101] HUANG J.S., KOTZ S., *Correlation structure in iterated Farlie-Gumbel-Morgenstern distributions*, Biometrika, **71**, 633-636, 1984.

[102] HUANG J.S., KOTZ S., *Modifications of the Farlie-Gumbel-Morgenstern distributions. A tough hill to climb*, Metrika, **49**, 307-323, 1999.

[103] HU T., *Negatively superadditive dependence of random variables with applications*, Dept of Statist. and Finance, University of Science and Technology, Hefei, China, 2000.

[104] HU T.,*Sufficient conditions for negative association of random variables*, Stat. and Prob. lett., **45**, 167-173, 1999.

[105] IRZIK G., *Can cause be reduce to correlations?*, Brit. J. Phil. Sci., **47**, 249-270, 1996.

[106] JOAG-DEV K., PROSCHAN F., *Negative Association of random variables*, Ann. Stat., **Vol 11** , 286-295, 1983.

[107] JOAG-DEV K., PERLMAN M., PITT L., *Association of normal random variables and Slepian's inequality*, Ann. Prob., **11**, 1983.

[108] JOE H., *Parametric Families of Mutivariate Distributions with given margins*, J. Multi. Anal. , **46**, 262-282, 1993.

[109] JOE H., *Majorization, randomness and dependence for multivariate distributions*, Ann. Prob., **15**, 3, 1227-1225, 1987.

[110] JOE H., *Relative entropy measures of multivariate dependence* , J. Amer. Statist. Assoc., **84**, 405, 157-164, 1989.

[111] JOE H., *Multivariate models and dependence concepts*, Chapman and Hall, 1997.

[112] JOHNSON N.L., KOTZ S., *On some generalized Farlie-Gumbel-Morgenstern Distributions*, Comm in Statist, **4**,(5), 415-428, 1975.

[113] JOHNSON N.L., KOTZ S. *A vector multivariate hazard rate*, J. Multi. Anal., **5**, 53-66, 1975.

[114] JOHNSON N.L., KOTZ S., *On some generalized Farlie-Gumbel-Morgenstern Distributions II*, Institute of Statistics, University of North Carolina, Mimeo series, 1080, July 1976; Comm in Statist., **A6**, 485-496, 1977.

[115] JOHNSON N.L., KOTZ S., *On an L_1 measure of dependence for modified bivariate uniform distributions*, Statistica (Bologna), **60**, 2000.

[116] JOHNSON N.L., KOTZ S., BALAKRISHNA N., *Continuous Univariate Distributions*, Vol. 2 (Second Edition), Chapter 32, New-York, Wiley, 1995.

[117] JOHNSON R.A., WICHERN D.W., *Applied Multivariate Analysis*, Third Edition, Prentice and Hall, Englewood, N.J., 1992.

[118] JONES M.C., *The local dependence function*, Biometrika , **83**, 4, 899-904, 1996.

[119] JONES M.C., *Constant local dependence*, J., Multi. Anal., **64**, 148-155, 1998.

[120] JOUINI M.N., CLEMEN R.T., *Copula models for aggregating expert opinions*, Oper. Research, **44**, n 3, 444-457, May-June 1996.

[121] KAPUR J.N., DHANDE M., *On the entropic measures of stochastic dependence*, Indian J. Pure Appl. Math., **17(5)**, 581-595, 1986.

[122] KALBFLEISCH J.D. , PRENTICE R.L., *The Statistical Analysis of Failure Time Data*, Wiley, 1980.

[123] KARLIN S., *Total Positivity*, Stanford University Press, 1968.

[124] KARLIN S., RINOTT Y.,*Classes of orderings of measures and related relations of inequalities*, J. Multi. Anal., **10**, 467-498, 1980.

[125] KEMPERMAN J.H.B., *On the FKG inequality measures on a partially ordered space*, Proc. Kon. Ned. Akad. Wet. Amsterdam, Series A, **80(4)**, 313-331, 1977.

[126] KEYNES J.M., *A Treatise on Probability*, London MacMillan, 1952, First edition 1921.

[127] KIMELDORF G., SAMPSON A.R., *One-parameter family of bivariate distributions with fixed marginals*, Comm. Statist., **4**, 293-301, 1975.

[128] KIMELDORF G., SAMPSON A.R., *Monotone dependence*, Ann. Stat., **6**, n4, 895-903, 1978.

[129] KIMELDORF G., SAMPSON A.R., *Positive dependence orderings*, Ann. Inst. Statist., **39**, Part A, 113-128, 1987.

[130] KIMELDORF G., SAMPSON A.R., *A framework for positive dependence*, Ann. Inst. Statist., **41**, n1, 31-45, 1989.

[131] KLAASSEN C.A., WELLNER J.A., *Efficient estimation in the bivariate normal copula model: normal margins are least favourable*, Bernoulli **3** (1), 55-77, 1997.

[132] KOTZ S., SEEGER J.P., *A new approach to dependence in multivariate distributions*, Advances in Probability Distributions with given marginals (G. Dall'Aglio *et al.*, Eds.), Dordrecht : Kluwer Academic Publishers, 113-127, University Press, 1991.

[133] KOTZ S., WANG Q.S., HUNG K., *Interrelations among various definitions of bivariate positive dependence*, Topics in Statistical Dependence, IMS Lecture Notes **16**, (H.W. Block, A.R. Sampson and T. Savits, Eds.), 1992.

[134] KOTZ S., SOONG C., *On Measures of dependence*, Technical Report, Temple University, Philadelphia, U.S.A., 1980.

[135] KOTZ S., BALAKRISHNAN N., JOHNSON N.L., *Multivariate continuous distributions*, Second Edition, Wiley, New York, 1972.

[136] KRAJKA A., SZYNAL D., *On Q-covariance and its applications*, Proceedings of International Conference on Linear Statistical Inference LINSTAT'93, (Caliński T. and Kala R. eds), Kluwer Academic Publishers, Dordrecht, 293-300, 1994.

[137] LACHENBRUCH P.A., BROGAN D.R., *Some distributions on the positive real line which have no moments*, Amer. Statist., February 1971.

[138] LAI C.D., XIE M., *A new family of positive quadrant dependent bivariate distributions*, Stat. Prob. Lett., **46**, 359-364, 2000.

[139] LAPLACE P.S., *Mémoire sur les intégrales définies et leurs applications aux probabilités*, Mem. Inst. Imp., France, 279-347, 1810.

[140] LEE M.L.T., *Properties and applications of the Sarmanov family of bivariate distributions*, Comm.Statist.-Theory Meth., **25** (6), 1207-1222, 1996.

[141] LEE L. ,*Multivariate distributions having Weibull properties* ,J. Multi. Anal., **9**, 267-277, 1979.

[142] LEHMANN E.L., *Some concepts of dependence*, Ann. Math. Stat., **37**, 1137-1153, 1966.

[143] LI H., SCARSINI M., SHAKED M., *Linkages : a tool for the construction of multivariate distributions with given nonoverlapping multivariate marginals*, J. Multi. Anal., **56**, 20-41, 1996.

[144] LING C.H., *Representation of associative functions* , Publ. Math. Debrecen, **12**, 189-222, 1965.

[145] LIN P., *Measures of association between vectors*, Commun. Statist.–Theory Meth, **16(2)**, 321-338, 1987.

[146] LONG D., KRZYSZTOFOWICZ R., *A family of bivariate densities constructed from marginals*, J. Amer. Stat. Assoc, **90**, n 430, 1995.

[147] LONG D., KRZYSZTOFOWICZ R., *Geometry of a correlation coeffi-*

cient under a copula , Commun. Statist.–Theory Meth., **25** (6), 1397-1404, 1996.

[148] MACKENZIE D.A., *Statistics in Britain : 1865-1930*, Edinburgh : Edinburgh University Press, 1981.

[149] MARCH H.L., *Comparaison numérique des courbes statistiques*, J. Soc. Statist. Paris, **46**, 255-277, 306-311, 1905.

[150] MARDIA K.,V., *The multivariate Pareto distribution*, Ann. Math. Stat., **33**, 1008-1015, 1962.

[151] MARSHALL A.W., OLKIN I., *Inequalities. Theory of majorization and its applications*, Academic Press, New-York, 1979.

[152] MARSHALL A.W., OLKIN I., *Families of multivariate Distributions*, J. Amer. Stat. Assoc, **83**, n 403, 834-841, 1988.

[153] MAYER-WOLF E. , *The Cramér-Rao functional and limiting laws* , Ann. Prob., **18**, 840-850, 1990.

[154] MEESTER S.G., MACKAY J., *A parametric model for cluster categorical data*, Biometrics, **50**, 954-963, 1994.

[155] METRY M.H., SAMPSON A.R., *A family of partial orderings for positive dependence among fixed marginal bivariate distributions*, Technical report, University of Pittsburgh, 1991.

[156] MORGENSTERN D., *Einfache Beispiele zweidimensionaler Verteilung*, Mitteislingsblatt für Mathematische Statistik, **8**, 234-235, 1956.

[157] MOSTELLER F., TUKEY J.W., *Data Analysis and Regression*, Reading, Mass.: Addison Wesley, 1977.

[158] MÜLLER A., *Stochastic orders generated by integrals: a unified study*, Adv. Appl. Prob., **29**, 414-428, 1997.

[159] MÜLLER A., SCARSINI M., *Stochastic comparison of random vectors with a common copula*, Technical Report WIOR 554, University of Karlsruhe, February 2000.

[160] NELSEN R.B., *On measures of association as measures of positive dependence*, Statist. Prob. Lett., 14, 269-274, 1992.

[161] NELSEN R.B., *An Introduction to copulas*, Springer-Verlag, New-York, 1998.

[162] NELSON W., *Hazard plotting for incomplete failure data*, J. Qual. Techn., **1**, 2-52, 1969.

[163] OAKES D., *Bivariate survival models induced by frailties*, J. Amer. Stat. Assoc., **84**, 406, 487-93, 1989.

[164] PAPATHANASSIOU V., *Some characteristic properties of the Fisher information matrix via Cacoullos-type inequalities*, J. Multi. Anal., **44**, 256-265, 1993.

[165] PEARSON K., *Contribution to the theory of evolution*, XIII, Draper's Co. Research memoirs, Biom. Series I., 1894.

[166] PEARSON K., *Notes on the history of statistics*, Biometrika, **13**, 25-45, 1920, (Reproduced in Studies in the History of Statistics, London: Griffin, 1970).

[167] PERSONS W.M., *The construction of a business barometer based upon annual data*, Amer. Econ. Rev, 739-769, 1916.

[168] PICKANDS J., *Multivariate extreme value distributions*, Bull. Int. Statist. Inst., 859-878, 1981.

[169] PLACKETT R.L., *A class of bivariate distributions*, J. Amer. Stat. Assoc., **60**, 517-522, 1965.

[170] PLANA G.A.A., *Mémoire sur divers problèmes de probabilité*, Mém. Acad. Turin, 1811.

[171] PLANA G.A.A., *Allgemeine Formeln um nach der Methode der kleinste Quadrate die Verbesserungen von 6 elementaren zu bestimmen*, Z. für Astron. verw. Wissen, **6**, 249-264, 1818.

[172] PLANA G.A.A., *Soluzione generale d'un problema di probabilita*, Mem. Mat. Fis. Soc. Ital., Modena, **18**, 31-45, 1820.

[173] PRENTICE R.L., CAI J., *Covariance and survivor function estimation using censored multivariate failure time data*, Biometrika, **79**, 3, 495-512, 1992.

[174] QUESADA-MOLINA J.J., *A generalization of an identity of Hoeffding and some applications*, J. Ital. Statist. Soc., 3, 405-411, 1992.

[175] QUETELET A., *Sur la probabilité de mesurer l'influence de causes qui modifient les éléments sociaux*, Corr. Math. et Publ., **7**, 321-346, 1832.

[176] RAO C.R., *Multivariate Analysis. Some reminiscences on its origin and development*, Sankyà, Ser. B, **45**, 284-299, 1983.

[177] REED W.G., *The correlation coefficient*, J. Amer. Stat. Assoc., **15**, 670-683, 1918.

[178] REIMANN J., *Positively quadrant dependent bivariate distributions with given marginals*, Periodica Polytechnica, Budapest, **32**, n 1-2, 1988.

[179] RENYI A., *On measures of dependence*, Acta Mathematica Academiae Scientarium Hungaricae, **10**, 441-451, 1959.

[180] RENYI A., *Quelques remarques sur les probabilités des évènements dépendants*, J. Math. Pures Appl., **9**, 393-398, 1937.

[181] REITZ H.L., *On functional relations for which the coefficient of correlation is zero*, J. Amer. Statist. Assoc., **16**, 472-476, 1919.

[182] RINOTT Y., POLLACK M., *A stochastic ordering induced by a concept of positive dependence and monotonicity of asymptotic test sizes* , Ann. Statist., **Vol 8**, 190-198 , 1980.

[183] RODGERS J.L., NICEWANDER W.A., *Thirteen ways to look at the correlation coefficient*, Amer. Statist., **42**, 59-66, 1988.

[184] ROUSSAS G.G., *Kernel estimates under association : strong uniform consticency*, Statist. Prob. Lett., **12**, 393-403, 1991.

[185] ROUSSAS G.G., *Positive and negative dependence with some statistical applications*, in Asymptotics Nonparametrics and Time Series, Ed. : S. Ghosh, New-York, M. Dekker, 1999.

[186] RÜSCHENDORF L., *Construction of Multivariate distributions with given marginals*, Ann. Inst. Statist. Math., **37**, Part A, 225-233, 1985.

[187] SARMANOV I.O., *New forms of correlation relationships between positive quantities applied in hydrology*, Math. Models in Hydrology Symposium. IAHS, **100**, 104-109, 1974.

[188] SARMANOV O.V., *Generalized normal correlation and two dimensional Fréchet classes*, Doklady AN SSSR, 168 (1), 32-35, 1966.

[189] SCARSINI M. SHAKED M., *Positive Dependence Orders: a survey* , Athens Conference in Applied Probability and Time Series 1 : Applied Probability Eds. : C. C. Heyde, Y.V. Prohorov , F. Pyke and S.T. Rachev, 1996.

[190] SCARSINI M. , *Multivariate stochastic dominance with fixed dependence structure*, Operations Research Letters, **7**, 237-240, 1988.

[191] SCARSINI M. VENETOULIAS A., *Bivariate distributions with non-monotone dependence structure*, J. Amer. Statist. Assoc., **88**, 338-344, 1993.

[192] SCARSINI M., *Multivariate convex orderings, dependence and stochastic equality*, J. Appl. Prob. , **35**, 93-103, 1998.

[193] SCHRIEVER B.F., *An ordering for positive dependence*, Ann. Statist., **15**, 1208-1214, 1987.

[194] SCHUCANY W., PARR W.C., BOYER J.E., *Correlation structure in Farlie-Gumbel-Morgenstern distributions*, Biometrika, **65**, 3, 650-653, 1978.

[195] SCHWARTZ L., *La théorie des distributions*, Hermann, Paris 1966.

[196] SCHWEIZER B., SKLAR A., *Probabilistic Metric Spaces*, North-Holland, New-York, 1983.

[197] SCHWEIZER B., WOLFF E.F., *On nonparametric measures of dependence for random variables*, Ann. Statist., **9**, 879-885, 1981.

[198] SEAL H.L., *The historical development of the Gauss linear model*, Biometrika, **54**, 1-24, 1987.

[199] SEARLE S.R., *Linear Models*, Wiley, N.Y., 1971.

[200] SHAKED M., *A note on the exchangeable generalized Farlie-Gumbel-Mogenstern distributions*, Comm. in Statist., **4**, 711-722, 1975.

[201] SHAKED M., *A family of concepts of dependence for bivariate distributions*, J. Amer. Statist. Assoc., **72**, 359, 642-650, 1977.

[202] SHAKED M., *A concept of positive dependence for exchangeable variables*, Ann. Statist., **5**, 105-115, 1977.

[203] SHAKED M., TONG Y.L., *Some partial orderings of exchangeable random variables* , J. Multi. Anal.,**17**, 333-349, 1985.

[204] SHAKED M., SHANTHIKUMAR J.G., *Stochastic orders and their applications*, Academic Press New-York 1995.

[205] SHAKED M., SHANTHIKUMAR G., *Supermodular stochastic orders and positive dependence of random vectors* , J. Multi. Anal., **61**, 86-101, 1997.

[206] SHIH J.H., LOUIS T.A., *Inferences on the association parameter in copula models for bivariate survival data*, Biometrics, **51**, 1384-1399, 1995.

[207] SHIH J.H., LOUIS T.A., *Tests of independence for bivariate survival data*, Biometrics **52**, 1440-1449, 1996.

[208] SKLAR M., *Fonctions de répartition à N dimensions et leurs marges*, Publ. Inst. Stat. Paris, **8**, 229-231, 1959.

[209] SLEPIAN D., *The one sided barrier problem for Gaussian noise*, Bell System Tech. J., **41**, 463-501, 1962.

[210] STIGLER S.M., *The History of Statistics*, Cambridge, MA : Belknap Press, 1986.

[211] STOYANOV J.M., *Counterexamples in probability* , Second edition, Wiley, New-York, 1997.

[212] SUNGUR E.A., YANG Y., *Diagonal copulas of archimedean class* , Commun. Statist. Theory Meth., **25** (7), 1659-1676, 1996.

[213] SUNGUR E.A., *Truncation invariant dependence structure*, Commun. Statist. Theory Meth., **28** (11), 2553-2568, 1999.

[214] THALIB L. BHATI M.I., *A significant correlation : does the sample size matter?*, Pak. J. Statist., **15** 2, 115-126, 1999.

[215] TAWN J.A., *Bivariate extreme value theory. Theory and estimation*, Biometrika, **75** 397-413, 1997.

[216] TCHEN A.H., *Inequalities for distributions with given marginals*, Ann. Prob., **8**, 814-827, 1980.

[217] THORNDIKE E.L., *Mental and Social Measurements*, New-York, 1913.

[218] TIAO G.C., GUTTMAN I.,*The Inverted Dirichlet distribution with applications*, J. Amer. Stat. Assoc., **60**, 793-805, and 1251-1252, 1965.

[219] TRÉGOÜET D-A., DUCIMETIÈRE P., BOCQUET V., VISVIKIS S., SOUBRIER F., TIRET L. , *A parametric copula model for analysis of familial binary data*, Amer. J. Hum. Genet., **64**, 886-893, 1999.

[220] TUNCER F., SUNGUR A.E., *Measures of positive and negative dependence*, The Frontiers of the Statistical Scientific Theory and Industrial Applications, Vol 2, Proceedings of First International Conference of Statistical Computing, Izmir, Turkey, 429-447, March-April 1987.

[221] TURNER D.W., *A simple example illustrating a well-known property of the correlation coefficient*, Amer. Statist., **51**, 1970.

[222] WALKER H.M., *Studies in the history of the statistical methods*, Baltimore MD: Williams and Wilkins,1929. Reprinted by Arno Press New-York, 1975.

[223] WEI G., FANG H.B., FANG K.T., *The dependence patterns of random variables - Elentary Algebraic and Geometrical properties of Copulas*, Technical Report, Hong-Kong Baptist University, 1998.

[224] WELDON W.F.R., *Certain correlation variations in Crangon vulgaris*, Proc. Roy. Soc. London, **51**, 2-21 1892.

[225] WOLFF E.F., *N-dimensional measures of dependence*, Stochastica, **4**, 175-188, 1980.

[226] YANAGIMOTO T., *Families of positively dependent random variables*, Ann. Inst. Statist., **24**, 559-573, 1972.

[227] YANAGIMOTO T., OKAMOTO M., *Partial orderings of permutations and monotonicity of a rank correlation statistic*, Ann. Inst. Statist., **21**, 489-506, 1969.

[228] YANAGIMOTO T., *Dependence ordering in statistical models and other notions*, Topics in Statistical Dependence , IMS Lecture Notes, H. W. Block *et al.* Eds., 1990.

[229] YASHIN A., IACHINE I.A., *How long can humans live. Mechanisms of ageing and development* **80**, 147-169, 1995.

[230] YASHIN A., IACHINE I.A., *Dependent hazards in multivariate survival problems*, J. Mult. Anal. **71**, 241-261, 1999.

[231] YULE G.U., *On the theory of Correlation*, J. Roy. Statist. Soc., **60**, 812-854, 1897.

[232] YULE G.U., *Introduction to the Theory of Statistics*, (Second Edition), London: Griffin, 1912.

[233] ZOGRAFOS K.,*On a measure of dependence based on Fisher's Information Matrix*, Commun. Statist.-Th. Meth., **27(7)**, 1715-1728, 1998.

[234] ZHENG M., KLEIN J.P., *A self-consistent estimator of marginal survival functions based on dependent competiting risk data and an assumed copula*, Comm. in Statist Theor. Meth., **23**(8), 2299-2311, 1994.

Index